Les E Kelly

Les E Kelly

Radiography in Modern Industry

FOURTH EDITION

EASTMAN KODAK COMPANY
Rochester, New York 14650

Acknowledgements

To W. R. Garrett, H. R. Splettstosser, D. E. Titus, and all of those who have contributed to this fourth edition of *Radiography in Modern Industry*, we extend our sincere appreciation.

Richard A. Quinn
Claire C. Sigl
Editors

John J. Callinan, Jr.
Director of Industrial Radiography

Contents

Introduction

Many of the spectacular scientific and engineering achievements of the past few years can be traced to nondestructive testing methods, which—by determining internal soundness without destroying product usefulness—assure the satisfactory performance for which the product was intended.

Radiography today is one of the most important, most versatile, of all the nondestructive test methods used by modern industry. Employing highly penetrating x-rays, gamma rays, and other forms of radiation that do not damage the part itself, radiography provides a permanent visible film record of internal conditions, containing the basic information by which soundness can be determined. In the past decade alone, the evidence from millions of film records, or radiographs, has enabled industry to assure product reliability; has provided the informational means of preventing accidents and saving lives; and has been beneficial for the user.

Since economic justification is a major criterion for any testing method, the value of radiography lies to some extent in its ability to make a profit for its user. This value is apparent in machining operations where only pieces known to be sound are permitted on the production lines. It is equally apparent in cost reductions when less expensive materials or fabricating methods can be employed instead of costlier ones in which soundness is only an estimated quality. The information gained from the use of radiography also assists the engineer in designing better products and protects the company by maintaining a uniform, high level of quality in its products. In total, these advantages can help to provide customer satisfaction and promote the manufacturer's reputation for excellence.

Industrial radiography is tremendously versatile. Objects radiographed range in size from micro-miniature electronic parts to mammoth missile components; in product composition through virtually every known material; and in manufactured form over an enormously wide variety of castings, weldments, and assemblies. Radiographic examination has been applied to organic and inorganic materials, and to solids, liquids, and even gases. An industry's production of radiographs may vary from the occasional examination of one or several pieces to the examination of hundreds of specimens per hour. This wide range of applications has resulted in the establishment of independent, professional x-ray laboratories as well as of radiographic departments within manufacturing plants themselves. The radiographic inspection performed by industry is frequently monitored for quality by its customers—other manufacturers or governmental agencies—who use, for the basis of monitoring, applicable specifications or codes, mutually agreed to by contract, and provided by several technical societies or other regulatory groups.

To meet the growing and changing demands of industry, research and development in the field of radiography are continually producing new sources of radiation such as neutron generators and radioactive isotopes; lighter, more powerful, more portable x-ray equipment as well as multimillion-volt x-ray machines designed to produce highly penetrating radiation; new and improved x-ray films and automatic film processors; and improved or specialized radiographic techniques. These factors, plus the activities of many dedicated people, extend radiography's usefulness to industry.

It is not surprising, then, that radiography, the first of the modern sophisticated methods of non-destructive testing (dating back to 1895), has led hundreds of industries to put great confidence in the information that it supplies. The list is growing year after year as industry's management, designers, engineers, production men, inspectors, and everyone concerned with sound practices, dependable products, high yields, and reasonable profits discover the value of radiography in modern industry.

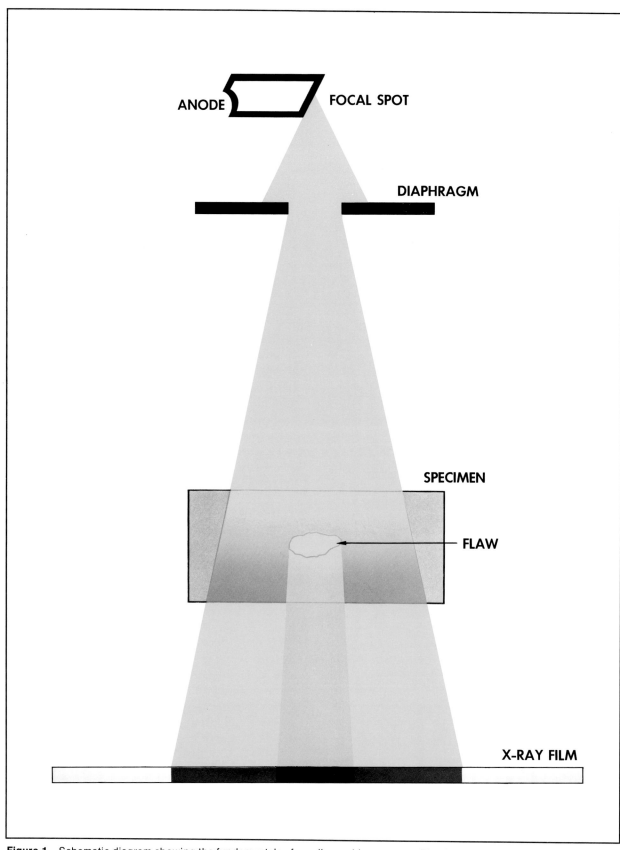

Figure 1—Schematic diagram showing the fundamentals of a radiographic exposure. The dark region of the film represents the more penetrable part of the object; the light regions, the more opaque.

The Radiographic Process

CHAPTER 1

NATURE OF X-RAYS

X-rays are a form of electromagnetic radiation (EMR), as is light. Their distinguishing feature is their extremely short wavelength—only about 1/10,000 that of light, or even less. This characteristic is responsible for the ability of x-rays to penetrate materials that absorb or reflect ordinary light.

X-rays exhibit all the properties of light, but in such a different degree as to modify greatly their practical behavior. For example, light is refracted by glass and, consequently, is capable of being focused by a lens in such instruments as cameras, microscopes, telescopes, and spectacles. X-rays are also refracted, but to such a very slight degree that the most refined experiments are required to detect this phenomenon. Hence, it is impractical to focus x-rays. It would be possible to illustrate the other similarities between x-rays and light but, for the most part, the effects produced are so different— particularly their penetration—that it is preferable to consider x-rays and gamma rays separately from other radiations. Figure 2 shows their location in the electromagnetic spectrum.

NATURE OF GAMMA RAYS

Gamma rays are similar in their characteristics to x-rays and show the same similarities to, and differences from, visible light as do x-rays. They are distinguished from x-rays only by their source, rather than by their nature. Gamma rays are emitted from the disintegrating nuclei of radioactive substances, and the quality (wavelength or penetration) and intensity of the radiation cannot be controlled by the user. Some gamma-ray-emitting radioactive isotopes, such as radium, occur naturally. Others, like cobalt 60, are artificially produced. In industrial radiography, the artificial radioactive isotopes are used almost exclusively as sources of gamma radiation.

MAKING A RADIOGRAPH

A *radiograph* is a photographic record produced by the passage of x-rays or gamma rays through an object onto a film (Figure 1). When film is exposed to x-rays, gamma rays, or light, an invisible change called a latent image is produced in the film emulsion. The areas so exposed become dark when the film is immersed in a developing solution, the degree of darkening depending on the amount of exposure. After development, the film is rinsed, preferably in a special bath, to stop development. The film is next put into a fixing bath, which dissolves the undarkened portions of the sensitive salt. It is then washed to remove the fixer and dried so that it may be handled, interpreted, and filed. The developing, fixing, and washing of the exposed film may be done either manually or in automated processing equipment.

The diagram in Figure 3 shows the essential features in the exposure of a radiograph. The focal spot is a small area in the x-ray tube from which the radiation emanates. In gamma radiography, it is the capsule containing the radioactive material, for

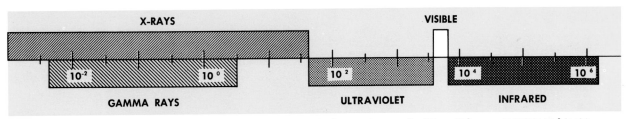

Figure 2—Portion of the electromagnetic spectrum. Wavelengths in angstrom units (1A = 10^{-8} cm = 3.937 X 10^{-9} inch).

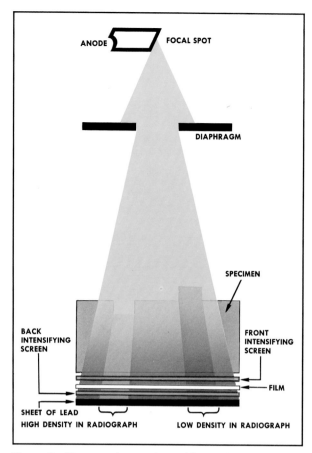

Figure 3—Diagram of setup for making an industrial radiograph with x-rays.

example, cobalt 60, that is the source of radiation. In either case the radiation proceeds in straight lines to the object; some of the rays pass through and others are absorbed—the amount transmitted depending on the nature of the material and its thickness. For example, if the object is a steel casting having a void formed by a gas bubble, the void results in a reduction of the total thickness of steel to be penetrated. Hence, more radiation will pass through the section containing the void than through the surrounding metal. A dark spot, corresponding to the projected position of the void, will appear on the film when it is developed. Thus, a radiograph is a kind of shadow picture—the darker regions on the film representing the more penetrable parts of the object, and the lighter regions, those more opaque to x- or gamma-radiation.

INTENSIFYING SCREENS

X-ray and other photographic films are sensitive to the direct action of the x-rays, but the photographic effect can be increased very appreciably, and exposure time can be decreased by the use of an inten-

sifying screen in contact with each side of the film.

One form of intensifying screen consists of lead foil, or a thin layer of a lead compound evenly coated on a paper backing. Under the excitation of x-rays of short wavelength and gamma rays, lead is a good emitter of electrons, which expose the sensitive film, thus increasing the total photographic effect.

Another form of intensifying screen consists of a powdered fluorescent chemical—for example, calcium tungstate, mixed with a suitable binder and coated on cardboard or plastic. Its action depends on the fact that it converts some of the x-ray energy into light, to which the film is very sensitive.

The decision as to the type of screen to be used—or whether a screen is to be used at all—depends on a variety of circumstances, which will be discussed in more detail later.

SCATTERED RADIATION

It is a property of *all materials* not only to absorb and transmit x-rays and gamma rays in varying degrees, but also to *scatter* them—as radiation of longer wavelength—in all directions. In radiography, the film receives scattered radiation from the object, the film holder, and any other material in the path of the primary x-ray beam. The effect is to diminish the contrast, detail, and clarity of the radiographic image. Lead screens, in contact with the film, lessen the relative effect of this longer-wavelength scattered radiation. Under some circumstances, a filter of copper or lead, placed between the x-ray tube and the object, or between the object and the film, diminishes the effect of scattered radiation upon the film. A lead mask that limits the volume of matter exposed to the primary radiation is sometimes helpful in lessening scatter. More detailed information on this subject is found in a later chapter.

TYPES OF FILM

Several special types of x-ray film have been designed for the radiography of materials. Some types work best with lead screens, or without screens. Other types are intended primarily for use with fluorescent intensifying screens. X-ray films are commonly coated with emulsion on both sides of the support—the superposition of the radiographic images of the two emulsion layers doubles the density and hence greatly increases the speed. X-ray films coated on one side only (single-coated films) are available for use when the superposed images in two emulsions might cause confusion.

X-ray and Gamma-ray Sources

PRODUCTION OF X-RAYS

X-rays are produced when electrons, traveling at high speed, collide with matter or change direction. In the usual type of x-ray tube, an incandescent filament supplies the electrons and thus forms the cathode, or negative electrode, of the tube. A high voltage applied to the tube drives the electrons to the anode, or target. The sudden stopping of these rapidly moving electrons in the surface of the target results in the generation of x-radiation.

The design and spacing of the electrodes and the degree of vacuum are such that no flow of electrical charge between cathode and anode is possible until the filament is heated.

THE X-RAY TUBE

Figure 4 is a schematic diagram of the essential parts of an x-ray tube. The filament is heated by a current of several amperes from a low-voltage source, generally a small transformer. The focusing cup serves to concentrate the stream of electrons on a small area of the target, called the focal spot. This stream of electrons constitutes the tube current and is measured in milliamperes.

The higher the temperature of the filament, the greater is its emission of electrons and the larger the resulting tube current. The tube current is controlled, therefore, by some device that regulates the heating current supplied to the filament. This is usually accomplished by a variable-voltage transformer which energizes the primary of the filament transformer. Other conditions remaining the same, the x-ray output is proportional to the tube current.

Most of the energy applied to the tube is trans-

formed into heat at the focal spot, only a small portion being transformed into x-rays. The high concentration of heat in a small area imposes a severe burden on the materials and design of the anode. The high melting point of tungsten makes it a very suitable material for the target of an x-ray tube. In addition, the efficiency of the target material in the production of x-rays is proportional to its atomic number.* Since tungsten has a high atomic number, it has a double advantage. The targets of practically all industrial x-ray machines are made of tungsten.

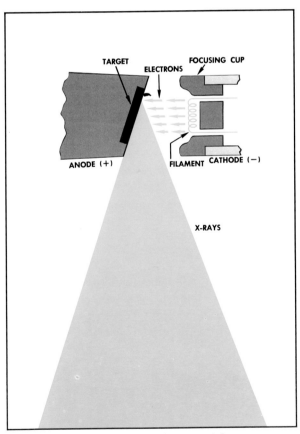

Figure 4—Schematic diagram of an x-ray tube.

---

*The atomic number of an element is the number of protons in the nucleus of the atom, and is equal to the number of electrons outside the nucleus. In the periodic table the elements are arranged in order of increasing atomic number. Hydrogen has an atomic number of 1; iron, of 26; copper, of 29; tungsten, of 74; and lead of 82.

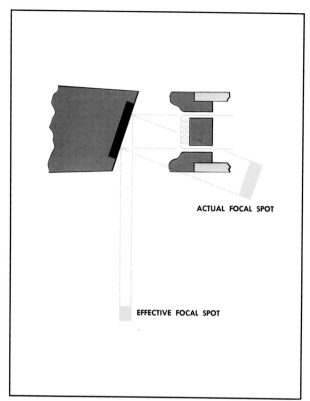

ACTUAL FOCAL SPOT

EFFECTIVE FOCAL SPOT

Figure 5—Diagram of a line-focus tube depicting relation between actual focal-spot area (area of bombardment) and effective focal spot, as projected from a 20° anode.

COOLING

Circulation of oil in the interior of the anode is an effective method of carrying away the heat. Where this method is not employed, the use of copper for the main body of the anode provides high heat conductivity, and radiating fins on the end of the anode outside the tube transfer the heat to the surrounding medium. The focal spot should be as small as conditions permit, in order to secure the sharpest possible definition in radiographic image. However, the smaller the focal spot, the less energy it will withstand without damage. Manufacturers of x-ray tubes furnish data in the form of charts indicating the kilovoltages and milliamperages that may be safely applied at various exposure times. The life of any tube will be shortened considerably if it is not always operated within the rated capacity.

FOCAL-SPOT SIZE

The principle of the line focus is used to provide a focal spot of small effective size, though the actual focal area on the anode face may be fairly large, as illustrated in Figure 5. By making the angle between the anode face and the central ray small, usually 20 degrees, the effective area of the spot is only a fraction of its actual area. With the focal area in the form of a long rectangle, the projected area in the direction of the central ray is square.

EFFECTS OF KILOVOLTAGE

As will be seen later, different voltages are applied to the x-ray tube to meet the demands of various classes of radiographic work. The higher the voltage, the greater the speed of the electrons striking the focal spot. The result is a decrease in the wavelength of the x-rays emitted and an increase in their penetrating power and intensity. It is to be noted that x-rays produced, for example, at 200 kilovolts contain all the wavelengths that would be produced at 100 kilovolts, and with greater intensity. In addition, the 200-kilovolt x-rays include some shorter wavelengths that do not exist in the 100-kilovolt spectrum at all. The higher voltage x-rays are used for the penetration of thicker and heavier materials.

Most x-ray generating apparatus consists of a filament supply and a high-voltage supply.

The power supply for the x-ray tube filament consists of an insulating *step-down* transformer. A variable-voltage transformer or a choke coil may serve for adjustment of the current supplied to the filament.

The high-voltage supply consists of a transformer, an autotransformer, and, quite frequently, a rectifier.

A transformer makes it possible to change the voltage of an alternating current. In the simplest form, it consists of two coils of insulated wire wound on an iron core. The coil connected to the source of alternating current is called the *primary* winding, the other the *secondary* winding. The voltages in the two coils are directly proportional to the number of turns, assuming 100 percent efficiency. If, for example, the primary has 100 turns, and the secondary has 100,000, the voltage in the secondary is 1,000 times as high as that in the primary. At the same time, the current in the coils is decreased in the same proportion as the voltage is increased. In the example given, therefore, the current in the secondary is only 1/1,000 that in the primary. A *step-up* transformer is used to supply the high voltage to the x-ray tube.

An autotransformer is a special type of transformer in which the output voltage is easily varied over a limited range. In an x-ray generator, the autotransformer permits adjustment of the primary voltage applied to the step-up transformer and, hence, of the high voltage applied to the x-ray tube.

The type of voltage *waveform* supplied by a high-voltage transformer is shown in Figure 6A and consists of alternating pulses, first in one direction and then in the other. Some industrial x-ray tubes are designed for the direct application of the high-voltage waveform of Figure 6A, the x-ray tube then acting as its own rectifier. Usually, however, the high voltage is supplied to a unit called a *rectifier*, which converts the pulses into the unidirectional form illustrated in Figure 6B. Another type of rectifier may convert the waveform to that shown in Figure 6C, but the general idea is the same in both cases—that is, unidirectional voltage is supplied to the x-ray tube. Sometimes a filter circuit is also provided which "smooths out" the voltage waves shown in Figure 6B and C, so that essentially constant potential is applied to the x-ray tube, Figure 6D. Many different high-voltage waveforms are possible, depending on the design of the x-ray machine and its installation. Figure 6 shows idealized waveforms difficult to achieve in practi-

cal high-voltage equipment. Departures from these forms may vary in different x-ray installations. Since x-ray output depends on the entire waveform, this accounts for the variation in radiographic results obtainable from two different x-ray machines operating at the same value of peak kilovoltage.

Tubes with the anodes at the end of a long extension cylinder are known as "rod-anode" tubes. The anodes of these tubes can be thrust through small openings (Figure 7, top) to facilitate certain types of inspection. If the target is perpendicular to the electron stream in the tube, the x-radiation through 360 degrees can be utilized (Figure 7, bottom), and an entire circumferential weld can be radiographed in a single exposure.

With tubes of this type, one special precaution is necessary. The long path of the electron stream down the anode cylinder makes the focusing of the electrons on the target very susceptible to magnetic influences. If the object being inspected is magnetized—for example, if it has undergone a magnetic inspection and has not been properly demagnetized—a large part of the electron stream can be

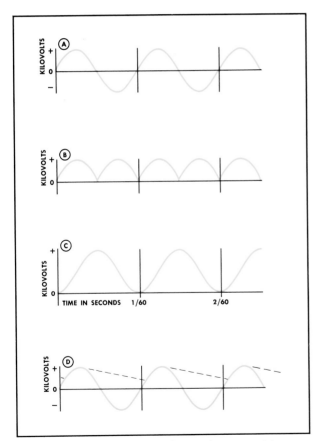

Figure 6—Typical voltage waveforms of x-ray machines.

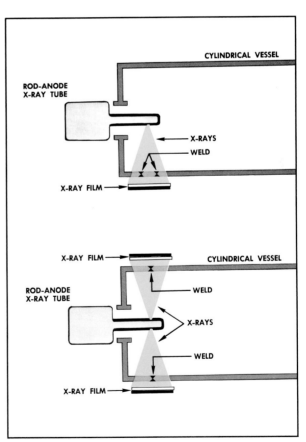

Figure 7—Top: Rod-anode tube used in the examination of a plug weld. **Bottom:** Rod-anode tube with a 360° beam used to examine a circumferential weld in a single exposure.

wasted on other than the focal-spot area, and the resulting exposures will be erratic.

The foregoing describes the operation of the most commonly used types of x-ray equipment. However, certain high-voltage generators operate on principles different from those discussed.

FLASH X-RAY MACHINES

Flash x-ray machines are designed to give extremely short (microsecond), extremely intense bursts of x-radiation. They are intended for the radiography of objects in rapid motion or the study of transient events (see page 117). The high-voltage generators of these units give a very short pulse of high voltage, commonly obtained by discharging a condenser across the primary of the high-voltage transformer. The x-ray tubes themselves usually do not have a filament. Rather, the cathode is so designed

that a high electrical field "pulls" electrons from the metal of the cathode by a process known as field emission, or cold emission. Momentary electron currents of hundreds or even thousands of amperes—far beyond the capacity of a heated filament—can be obtained by this process.

HIGH-VOLTAGE EQUIPMENT

The *betatron* may be considered as a high-voltage transformer, in which the secondary consists of electrons circulating in a doughnut-shaped vacuum tube placed between the poles of an alternating current electromagnet that forms the primary. The circulating electrons, accelerated to high speed by the changing magnetic field of the primary, are caused to impinge on a target within the accelerating tube.

In the *linear accelerator*, the electrons are acceler-

TABLE I—TYPICAL X-RAY MACHINES AND THEIR APPLICATIONS

Maximum Voltage (kV)	Screens	Applications and Approximate Thickness Limits
50	None	Thin sections of most metals; moderate thickness of graphite and beryllium; small electronic components; wood, plastics, etc.
150	None or lead foil	5-inch aluminum or equivalent. (See Table IV.) 1-inch steel or equivalent.
150	Fluorescent	1½-inch steel or equivalent. (See Table IV.)
300	Lead foil	3-inch steel or equivalent.
300	Fluorescent	4-inch steel or equivalent.
400	Lead foil	3½-inch steel or equivalent.
400	Fluorescent	4½-inch steel or equivalent.
1000	Lead foil	5-inch steel or equivalent.
1000	Fluorescent	8-inch steel or equivalent.
2000	Lead foil	8-inch steel or equivalent.
8 to 25 MeV	Lead foil	16-inch steel or equivalent.
8 to 25 MeV	Fluorescent	20-inch steel or equivalent.

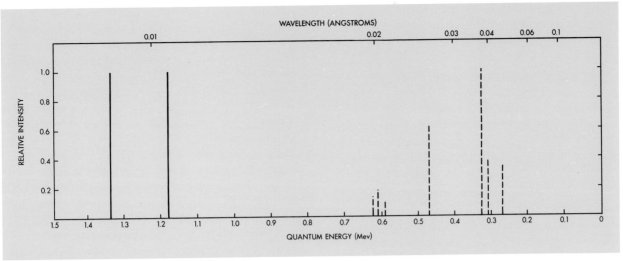

Figure 8—Gamma-ray spectrum of cobalt 60 (solid lines) and principal gamma rays of iridium 192 (dashed lines).

ated to high velocities by means of a high-frequency electrical wave that travels along the tube through which the electrons travel.

Both the betatron and the linear accelerator are used for the generation of x-radiation in the multimillion-volt range.

In the high-voltage *electrostatic generator*, the high voltage is supplied by static negative charges mechanically conveyed to an insulating electrode by a moving belt. Electrostatic generators are used for machines in the 1- and 2-million-volt range.

No attempt is made here to discuss in detail the various forms of electrical generating equipment. The essential fact is that electrons must be accelerated to very great velocities in order that their deceleration, when they strike the target, may produce x-radiation.

In developing suitable exposure techniques, it is important to know the voltage applied to the x-ray tube. It is common practice for manufacturers of x-ray equipment to calibrate their machines at the factory. Thus, the operator may know the voltage across the x-ray tube from the readings of the voltmeter connected to the primary winding of the high-voltage transformer.

APPLICATION OF VARIOUS TYPES OF X-RAY APPARATUS

The various x-ray machines commercially available may be roughly classified according to their maximum voltage. The choice among the various classes will depend on the type of work to be done. Table I lists voltage ranges and applications of typical x-ray machines. The voltage ranges are approximate since the exact voltage limits of machines vary from one manufacturer to another. It should be emphasized that a table of the type of Table I can serve only as the roughest sort of guide, since x-ray machines differ in their specifications, and radiographic tasks differ in their requirements.

X-ray machines may be either fixed or mobile, depending on the specific uses for which they are intended. When the material to be radiographed is portable, the x-ray machine is usually permanently located in a room protected against the escape of x-radiation. The x-ray tube itself is frequently mounted on a stand allowing considerable freedom of movement. For the examination of objects that are fixed or that are movable only with great difficulty, mobile x-ray machines may be used. These may be truck-mounted for movement to various parts of a plant, or they may be small and light enough to be carried onto scaffolding, through manholes, or even self-propelled to pass through pipelines. Semiautomatic machines have been designed for the radiography of large numbers of relatively small parts on a "production line" basis. During the course of an exposure, the operator may arrange the parts to be radiographed at the next exposure, and remove those just radiographed, with an obvious saving in time.

GAMMA-RAY SOURCES

Radiography with gamma rays has the advantages of simplicity of the apparatus used, compactness of the radiation source, and independence from outside power. This facilitates the examination of pipe, pressure vessels, and other assemblies in which access to the interior is difficult; field radiography of structures remote from power supplies; and radiography in confined spaces, as on shipboard.

In contradistinction to x-ray machines, which emit a broad band of wavelengths (see page 22), gamma-ray sources emit one or a few discrete wavelengths. Figure 8 shows the gamma-ray spectrum of cobalt 60 and the principal gamma rays of iridium 192. (The most intense line in each spectrum has been assigned an intensity of 1.0.)

Note that gamma rays are most often specified in terms of the energy of the individual photon, rather than in the wavelength. The unit of energy used is the electron volt (eV)—an amount of energy equal to the kinetic energy an electron attains in falling through a potential difference of 1 volt. For gamma rays, multiples—kiloelectron volts (keV; 1 keV = 1,000 eV) or million electron volts (MeV; 1 MeV = 1,000,000 eV)—are commonly used. A gamma ray with an energy of 0.5 MeV (500 keV) is equivalent in wavelength and in penetrating power to the most penetrating radiation emitted by an x-ray tube operating at 500 kV. The bulk of the radiation emitted by such an x-ray tube will be much less penetrating (much softer) than this (see Figure 18). Thus the radiations from cobalt 60, for example, with energies of 1.17 and 1.33 MeV, will have a penetrating power (hardness) about equal to that of the radiation from a 2-million-volt x-ray machine.

For comparison, a gamma ray having an energy of 1.2 MeV has a wavelength of about 0.01 angstrom (A); a 120 keV gamma ray has a wavelength of about 0.1 angstrom.

The wavelengths (or energies of radiation) emitted by a gamma-ray source, and their relative intensities, depend *only* on the nature of the emitter. Thus, the radiation *quality* of a gamma-ray source is not variable at the will of the operator.

The gamma rays from cobalt 60 have relatively great penetrating power and can be used, under some conditions, to radiograph sections of steel 9 inches thick, or the equivalent. Radiations from other radioactive materials have lower energies; for example, iridium 192 emits radiations roughly equivalent to the x-rays emitted by a conventional x-ray tube operating at about 600 kV.

The *intensity* of gamma radiation depends on the strength of the particular source used—specifically, on the number of radioactive atoms in the source that disintegrate in one second. This, in turn, is usually given in terms of curies.* For small- or moderate-sized sources emitting penetrating gamma rays, the intensity of radiation emitted from the source is proportional to the source activity in curies. The proportionality between the external gamma-ray intensity and the number of curies fails, however, for large sources or for those emitting relatively low-energy gamma rays. In these latter cases, gamma radiation given off by atoms in the middle of the source will be appreciably absorbed *(self-absorption)* by the overlying radioactive material itself. Thus, the intensity of the useful radiation will be reduced to some value below that which would be calculated from the number of curies and the radiation output of a physically small gamma-ray source.

A term often used in speaking of radioactive sources is *specific activity*, a measure of the degree of concentration of a radioactive source. Specific activity is usually expressed in terms of curies per gram. Of two gamma-ray sources of the same material and activity, the one having the greater specific activity will be the smaller in actual physical size. Thus, the source of higher specific activity will suffer less from self-absorption of its own gamma radiation. In addition, it will give less geometrical unsharpness in the radiograph or, alternatively, will allow shorter source-film distances and shorter exposures (see page 15).

Gamma-ray sources gradually lose activity with time, the rate of decrease of activity depending on the kind of radioactive material (see Table II). For instance, the intensity of the radiation from a cobalt 60 source decreases to half its original value in about 5 years; and that of an iridium 192 source, in about 70 days. Except in the case of radium, now little used in industrial radiography, this decrease in emission necessitates more or less frequent revision of exposures and replacement of sources.

The exposure calculations necessitated by the gradual decrease in the radiation output of a gamma-ray source can be facilitated by the use of decay curves similar to those for iridium 192 shown in Figure 9. The curves contain the same information, the only difference being that Figure 9 (left) shows activity on a linear scale, and Figure 9 (right), on a logarithmic scale. The type shown in Figure 9 (right) is easiest to draw. Locate point X, at the intersection of the half-life of the isotope (horizontal scale) and the "50 percent remaining activity" line (vertical scale). Then draw a straight line from the "zero time, 100 percent activity" point Y through point X.

---

*Curie—1 Ci = 3.7 x 10^{10}s^{-1}.

TABLE II—RADIOACTIVE MATERIALS USED IN INDUSTRIAL RADIOGRAPHY

Radioactive Element	Half-Life	Energy of Gamma Rays (MeV)	Gamma-Ray Dosage Rate (roentgens* per hour per curie at 1 meter)
Thulium 170	127 days	0.084 and 0.54†	. . .
Iridium 192	70 days	0.137 to 0.651‡	0.55
Cesium 137	33 years	0.66	0.39
Cobalt 60	5.3 years	1.17 and 1.33	1.35

*The roentgen (R) is a special unit for x- and gamma-ray exposure (ionization of air): 1 roentgen = 2.58 x.10⁻⁴ coulombs per kilogram (Ckg⁻¹). (The International Commission on Radiation Units and Measurements [ICRU] recommends that the roentgen be replaced gradually by the SI unit [Ckg⁻¹] by about 1985.)

†These gamma rays are accompanied by a more or less intense background of much harder radiation. The proportion of hard radiation depends upon the chemical nature and physical size of the source.

‡Twelve gamma rays.

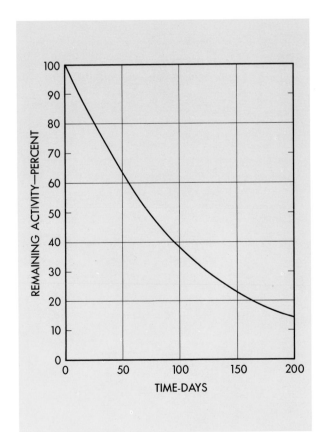
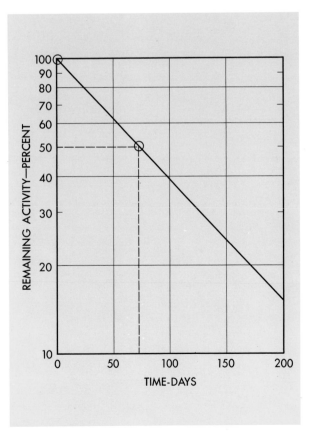

Figure 9—Decay curves for iridium 192. **Left:** Linear plot. **Right:** Logarithmic plot.

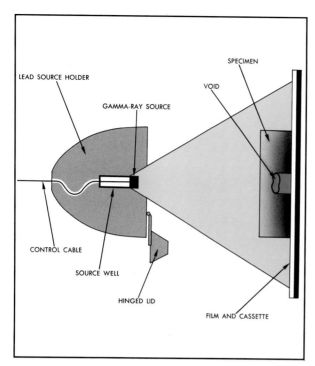

Figure 10—Typical industrial gamma-ray arrangement. Gamma-ray source in a combination "camera" and storage container.

It is difficult to give specific recommendations on the choices of gamma-ray emitter and source strength (Figure 10). These choices will depend on several factors, among which are the type of specimen radiographed, allowable exposure time, storage facilities available, protective measures required, and convenience of source replacement. The values given in Table III for practical application are therefore intended only as a rough guide and in any particular case will depend on the source size used and the requirements of the operation.

TABLE III—INDUSTRIAL GAMMA-RAY SOURCES AND THEIR APPLICATIONS

Source	Applications and Approximate Practical Thickness Limits
Thulium 170	Plastics, wood, light alloys. ½-inch steel or equivalent.
Iridium 192	1½ to 2½-inch steel or equivalent.
Cesium 137	1 to 3½-inch steel or equivalent.
Cobalt 60	2½ to 9-inch steel or equivalent.

CHAPTER 3

Geometric Principles

A radiograph is a shadow picture of an object that has been placed in the path of an x-ray or gamma-ray beam, between the tube anode and the film or between the source of gamma radiation and the film. It naturally follows, therefore, that the appearance of an image thus recorded is materially influenced by the relative positions of the object and the film and by the direction of the beam. For these reasons, familiarity with the elementary principles of shadow formation is important to those making and interpreting radiographs.

GENERAL PRINCIPLES

Since x-rays and gamma rays obey the common laws of light, their shadow formation may be explained in a simple manner in terms of light. It should be borne in mind that the analogy between light and these radiations is not perfect since all objects are, to a greater or lesser degree, transparent to x-rays and gamma rays and since scattering presents greater problems in radiography than in optics. However, the same geometric laws of shadow formation hold for both light and penetrating radiation.

Suppose, as in Figure 11A, that there is light from a point L falling on a white card C, and that an opaque object O is interposed between the light source and the card. A shadow of the object will be formed on the surface of the card.

This shadow cast by the object will naturally show some *enlargement* because the object is not in contact with the card; the *degree of enlargement* will vary according to the relative distances of the object from the card and from the light source. The law governing the size of the shadow may be stated:

The diameter of the object is to the diameter of the shadow as the distance of the light from the object is to the distance of the light from the card.

Mathematically, the degree of enlargement may be calculated by use of the following equations:

$$\frac{S_o}{S_i} = \frac{D_o}{D_i} \text{ or } S_o = S_i \left(\frac{D_o}{D_i}\right)$$

where S_o is the size of the object; S_i is the size of the shadow (or the radiographic image); D_o is the distance from source of radiation to object; and D_i the distance from the source of radiation to the recording surface (or radiographic film).

The degree of *sharpness* of any shadow depends on the size of the source of light and on the position of the object between the light and the card—whether nearer to or farther from one or the other. When the source of light is not a point but a small area, the shadows cast are not perfectly sharp (Figure 11, B to D) because *each point* in the source of light casts its own shadow of the object, and each of these overlapping shadows is slightly displaced from the others, producing an ill-defined image.

The *form of the shadow* may also differ according to the angle that the object makes with the incident light rays. Deviations from the *true shape* of the object as exhibited in its shadow image are referred to as distortion.

Figure 11, A to F shows the effect of changing the size of the source and of changing the relative positions of source, object, and card. From an examination of these drawings, it will be seen that the following conditions must be fulfilled to produce the sharpest, truest shadow of the object:

1. The source of light should be small, that is, as nearly a point as can be obtained. Compare Figure 11, A and C.

2. The source of light should be as far from the object as practical. Compare Figure 11, B and C.

3. The recording surface should be as close to the object as possible. Compare Figure 11, B and D.

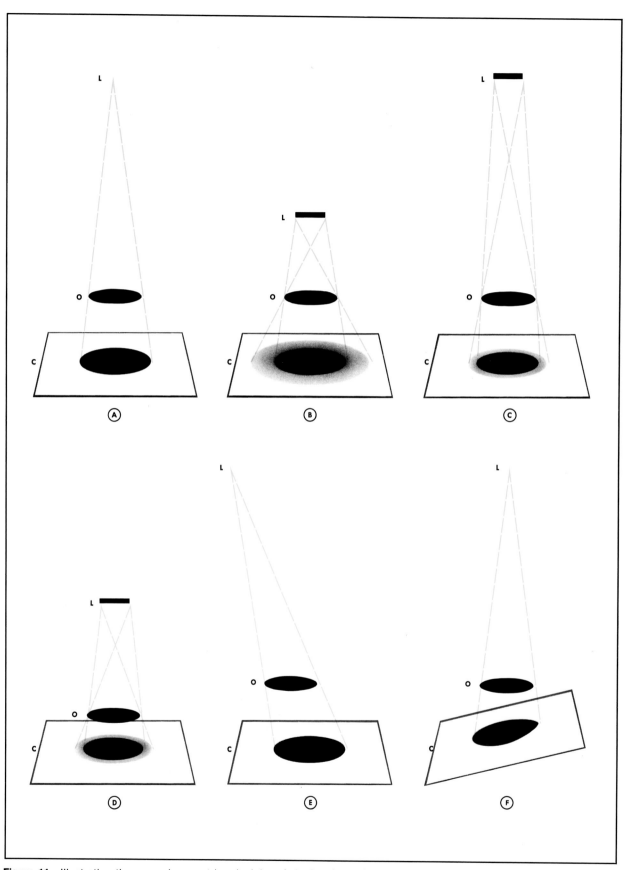

Figure 11—Illustrating the general geometric principles of shadow formation as explained on pages 15 to 19.

4. The light rays should be directed perpendicularly to the recording surface. See Figure 11, A and E.

5. The plane of the object and the plane of the recording surface should be parallel. Compare Figure 11, A and F.

RADIOGRAPHIC SHADOWS

The basic principles of shadow formation must be given primary consideration in order to assure satisfactory sharpness in the radiographic image and essential freedom from distortion. A certain degree of distortion naturally will exist in every radiograph because some parts will always be farther from the film than others, the greatest magnification being evident in the images of those parts at the greatest distance from the recording surface (see Figure 11).

Note, also, that there is no distortion of shape in Figure 11E—a circular object having been rendered as a circular shadow. However, under circumstances similar to those shown in Figure 11E, it is possible that spatial relations can be distorted. In Figure 12 the two circular objects can be rendered

either as two circles (Figure 12A) or as a figure-eight-shaped shadow (Figure 12B). It should be observed that both lobes of the figure eight have circular outlines.

Distortion cannot be eliminated entirely, but by the use of an appropriate source-film distance, it can be lessened to a point where it will not be objectionable in the radiographic image.

APPLICATION TO RADIOGRAPHY

The application of the geometric principles of shadow formation to radiography leads to five general rules. Although these rules are stated in terms of radiography with x-rays, they also apply to gamma-ray radiography.

1. The focal spot should be as small as other considerations will allow, for there is a definite relation between the size of the focal spot of the x-ray tube and the *definition* in the radiograph. A large-focus tube, although capable of withstanding large loads, does not permit the delineation of as much detail as a small-focus tube. Long

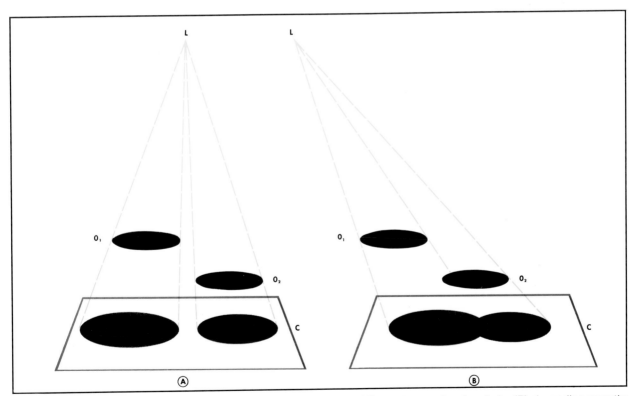

Figure 12—Two circular objects can be rendered as two separate circles **(A)** or as two overlapping circles **(B)**, depending upon the direction of the radiation.

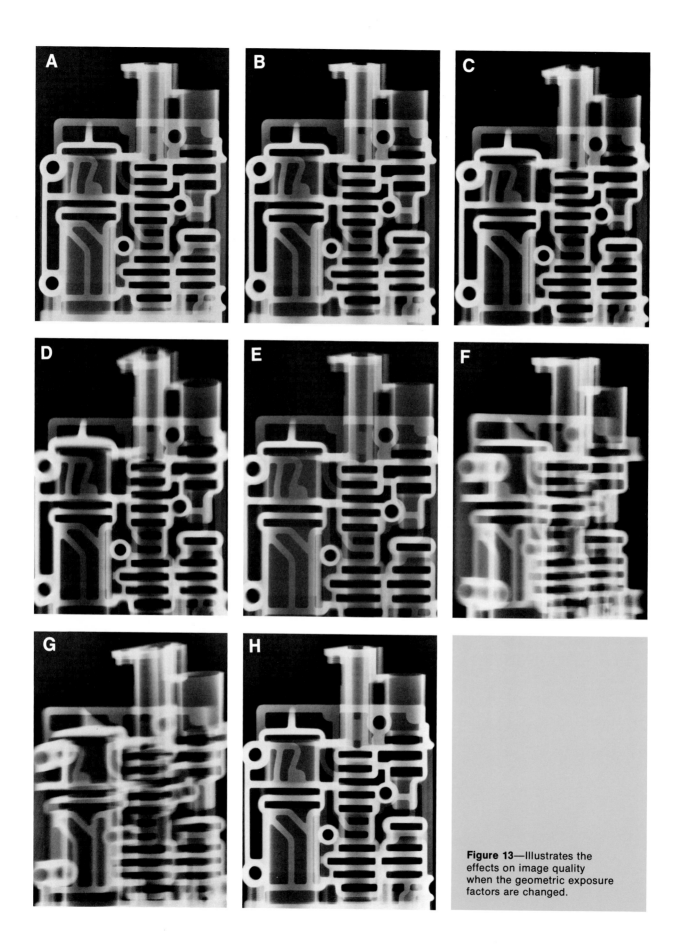

Figure 13—Illustrates the effects on image quality when the geometric exposure factors are changed.

source-film distances will aid in showing detail when a large-focus tube is employed, but it is advantageous to use the smallest focal spot permissible for the exposures required.

Figure 13, B and H show the effect of focal spot size on image quality. As the focal spot size is increased from 1.5 mm (B) to 4.0 mm (H), the definition of the radiograph starts to degrade. This is especially evident at the edges of the chambers which are no longer sharp.

2. The distance between the anode and the material examined should always be as great as is practical. Comparatively long source-film distances should be used in the radiography of thick materials to minimize the fact that structures farthest from the film are less sharply recorded than those nearer to it. At long distances, radiographic definition is improved and the image is more nearly the actual size of the object.

Figure 13, A to D show the effects of source-film distance on image quality. As the source-film distance is decreased from 68 inches (A) to 12 inches (D) the image becomes more distorted until at 12 inches it is no longer a true representation of the casting. This is particularly evident at the edges of the casting where the distortion is greatest.

3. The film should be as close as possible to the object being radiographed. In practice, the film —in its cassette or exposure holder—is placed in contact with the object.

In Figure 13, B and E the effects of object-film distance are evident. As the object-film distance is increased from zero (B) to 4 inches (E), the image becomes larger and the definition begins to degrade. Again, this is especially evident at the edges of the chambers which are no longer sharp.

4. The central ray should be as nearly perpendicular to the film as possible to preserve spatial relations.

5. As far as the shape of the specimen will allow, the plane of maximum interest should be parallel to the plane of the film.

Finally, in Figure 13, F and G the effects of object-film-source orientation are shown. When compared to B, the image F is extremely distorted because although the film is perpendicular to the central ray, the casting is at a 45° angle to the film and spatial relationships are lost. As the film is rotated to be parallel with the casting (G), the spatial relationships are maintained and the distortion is lessened.

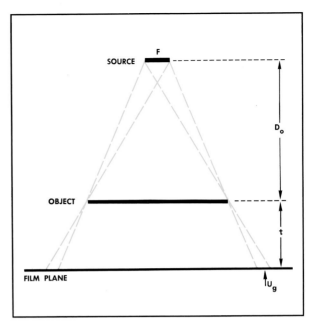

Figure 14—Geometric construction for determining geometric unsharpness (U_g).

CALCULATION OF GEOMETRIC UNSHARPNESS

The width of the "fuzzy" boundary of the shadows in Figures 11B, C, and D is known as the *geometric unsharpness* (U_g). Since the geometric unsharpness can strongly affect the appearance of the radiographic image, it is frequently necessary to determine its magnitude.

From the laws of similar triangles, it can be seen (Figure 14) that:

$$\frac{U_g}{F} = \frac{t}{D_o} \text{ or } U_g = F\left(\frac{t}{D_o}\right)$$

where U_g is the geometric unsharpness, F is the size of the radiation source, D_o is the source-object distance, and t is the object-film distance. Since the maximum unsharpness involved in any radiographic procedure is usually the significant quantity, the object-film distance (t) is usually taken as the distance from the *source side* of the specimen to the film.

D_o and t must be measured in the *same units;* inches are customary, but any other unit of length—say, centimetres—would also be satisfactory. So long as D_o and t are in the same units, the formula above will always give the geometric unsharpness U_g in whatever units were used to measure the dimensions of the source. The projected size of the focal spots of x-ray tubes (see Figure 5)

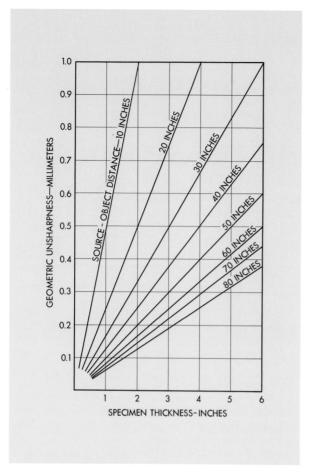

Figure 15—Graph relating geometric unsharpness (U_g) to specimen thickness and source-object distance, for a 5-millimetre source size.

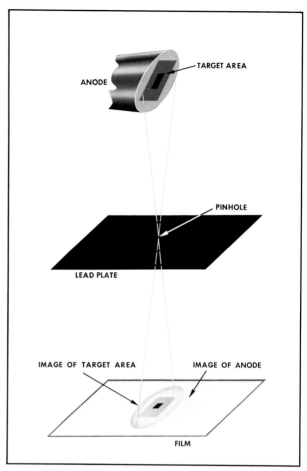

Figure 16—Schematic diagram showing production of a pinhole picture of an x-ray tube focal spot.

are usually stated in millimetres, and U_g will also be in millimetres. If the source size is stated in inches, U_g will be in inches.

For rapid reference, graphs of the type shown in Figure 15 can be prepared by the use of the equation on page 19. These graphs relate source-film distance, object-film distance and geometric unsharpness. Note that the lines of Figure 15 are all straight. Therefore, for each source-object distance, it is only necessary to calculate the value of U_g for a single specimen thickness, and then draw a straight line through the point so determined and the origin. It should be emphasized, however, that a separate graph of the type shown in Figure 15 must be prepared for each size of source.

PINHOLE PROJECTION OF FOCAL SPOT

Since the dimensions of the radiation source have considerable effect on the sharpness of the shadows, it is frequently desirable to determine the shape and size of the x-ray tube focal spot. This may be accomplished by the method of pinhole radiography, which is identical in principle with that of the pinhole camera. A thin lead plate containing a small hole is placed exactly midway between the focal spot and the film, and lead shielding is so arranged that no x-rays except those passing through the pinhole reach the film (Figure 16). The developed film will show an image that, for most practical radiographic purposes, may be taken as equal in size and shape to the focal spot (Figure 17). If precise measurements are required, the measured dimensions of the focal-spot image should be decreased by twice the diameter of the pinhole.

The method is applicable to x-ray tubes operating up to about 250 kV. Above this kilovoltage, however, the thickness of the lead needed makes the method impractical. (The entire focal spot cannot be "seen" from the film side of a small hole in a thick plate.) Thus the technique cannot be used for high-energy x-rays or the commonly used gamma-ray sources, and much more complicated methods, suitable only for the laboratory, must be employed.

A focus-film distance of 24 inches is usually convenient. Of course, the time of exposure will be much greater than that required to expose the film without the pinhole plate because so little radiation can get through such a small aperture. In general, a needle or a No. 60 drill will make a hole small enough for practical purposes.

A density in the image area of 1.0 to 2.0 is satisfactory. If the focal-spot area is overexposed, the estimate of focal-spot size will be exaggerated, as can be seen by comparing the two images in Figure 17.

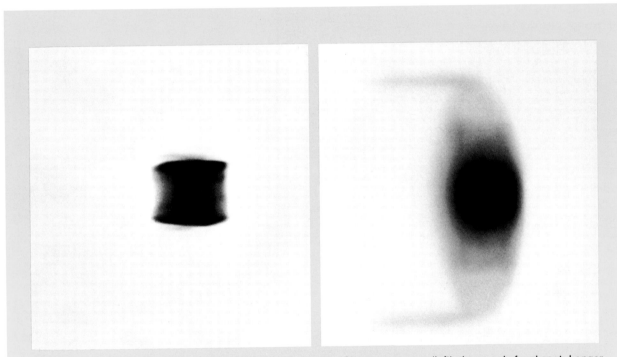

Figure 17—Pinhole pictures of the focal spot of an x-ray tube. Shorter exposure (left) shows only focal spot. Longer exposure (right) shows, as well as the focal spot, some details of tungsten button and copper anode stem. The x-ray images of these parts result from their bombardment with stray electrons.

CHAPTER 4

Factors Governing Exposure

Generally speaking, the density of any radiographic image depends on the amount of radiation absorbed by the sensitive emulsion of the film. This amount of radiation in turn depends on several factors: the total amount of radiation emitted by the x-ray tube or gamma-ray source; the amount of radiation reaching the specimen; the proportion of this radiation that passes through the specimen; and the intensifying action of the screens, if they are used. The effects of these factors will be discussed in this chapter.

RADIATION EMITTED BY SOURCE

X-rays

The total amount of radiation emitted by an x-ray tube depends on tube current (milliamperage), kilovoltage, and the time the tube is energized.

When other operating conditions are held constant, a change in milliamperage causes a change in the *intensity* of the radiation emitted, the intensity being approximately proportional to the milliamperage. The high voltage transformer saturation and voltage waveform can change with tube current, but a compensation factor is usually applied to minimize the effects of these changes. In normal industrial radiographic practice, the variation from exact proportionality is not serious and may usually be ignored.

Figure 18 shows spectral emission curves for an x-ray tube operated at two different currents, the higher being twice the lower. Therefore, *each wavelength* is twice as intense in one beam as in the other. Note that no wavelengths are present in one beam that are not present in the other. Hence, there is no change in x-ray quality or penetrating power.

As would be expected, the total amount of radiation emitted by an x-ray tube operating at a certain kilovoltage and milliamperage is directly proportional to the time the tube is energized.

Since the x-ray output is directly proportional to

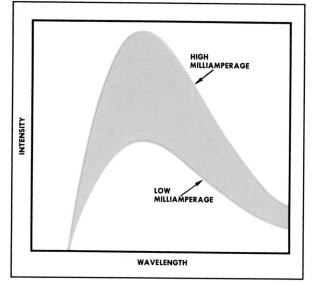

Figure 18—Curves illustrating the effect of a change in milliamperage on the intensity of an x-ray beam. (After Ulrey.)

both milliamperage and time, it is directly proportional to their product. (This product is often referred to as the "exposure.") Algebraically, this may be stated $E = Mt$, where E is the exposure, M the tube current, and t the exposure time. Hence, the amount of radiation will remain constant if the exposure remains constant, no matter how the individual factors of tube current and exposure time are varied. This permits specifying x-ray exposures in terms of milliampere-minutes or milliampere-seconds, without stating the specific individual values of tube current and time.

The kilovoltage applied to the x-ray tube affects not only the quality but also the intensity of the beam. As the kilovoltage is raised, x-rays of shorter wavelength, and hence of more penetrating power, are produced. Figure 19 shows spectral emission curves for an x-ray tube operated at two different kilovoltages but at the same milliamperage. Note that, in the higher-kilovoltage beam, there are some shorter wavelengths that are absent from the lower-kilovoltage beam. Further, all wavelengths present in the lower-kilovoltage beam are present in the more penetrating beam, and in greater amount. Thus, raising the kilovoltage increases both the penetration and the intensity of the radiation emitted from the tube.

Gamma Rays

The total amount of radiation emitted from a gamma-ray source during a radiographic exposure depends on the activity of the source (usually stated in curies) and the time of exposure. For a particular radioactive isotope, the intensity of the radiation is

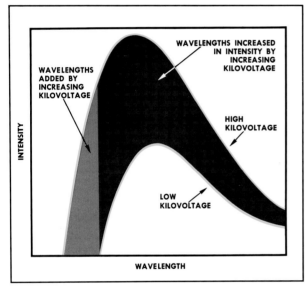

Figure 19—Curves illustrating the effect of a change in kilovoltage on the composition and intensity of an x-ray beam. (After Ulrey.)

approximately proportional to the activity (in curies) of the source. If it were not for absorption of gamma rays within the radioactive material itself (self-absorption, see page 12), this proportionality would be exact. In normal radiographic practice, the range of source sizes used in a particular location is small enough so that variations from exact proportionality are not serious and may usually be ignored.

Thus, the gamma-ray output is directly proportional to both activity of the source and time, and hence is directly proportional to their product. Analogously to the x-ray exposure, the gamma-ray exposure E may be stated $E = Mt$, where M is the

source activity in curies and t is the exposure time, the amount of gamma radiation remaining constant so long as the product of source activity and time remains constant. This permits specifying gamma-ray exposures in curie-hours without stating specific values for source activity or time.

Since gamma-ray energy is fixed by the nature of the particular radioactive isotope, there is no variable to correspond to the kilovoltage factor encountered in x-radiography. The only way to change penetrating power when using gamma rays is to change the source, i.e., cobalt 60 in place of iridium 192. See Figure 8, page 11.

INVERSE SQUARE LAW

When the x-ray tube output is held constant, or when a particular radioactive source is used, the radiation intensity reaching the specimen is governed by the distance between the tube (or source) and the specimen, varying inversely with the square of this distance. The explanation that follows is in terms of x-rays and light, but it applies to gamma rays as well.

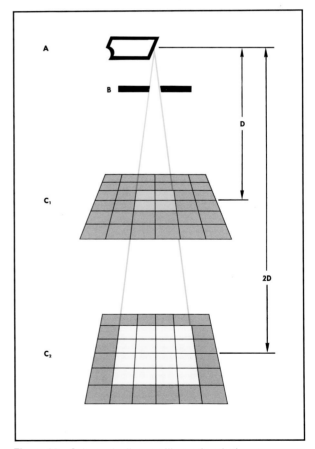

Figure 20—Schematic diagram illustrating the inverse square law.

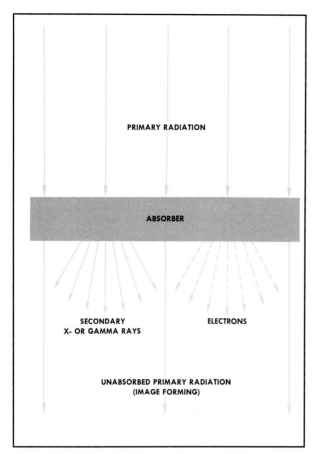

Figure 21—Schematic diagram of some of the ways x- or gamma-ray energy is dissipated on passing through matter. Electrons from specimens are usually unimportant radiographically; those from lead foil screens are very important.

PRIMARY RADIATION

ABSORBER

SECONDARY
X- OR GAMMA RAYS

ELECTRONS

UNABSORBED PRIMARY RADIATION
(IMAGE FORMING)

Since x-rays conform to the laws of light, they diverge when they are emitted from the anode and cover an increasingly larger area with lessened intensity as they travel from their source. This principle is illustrated in Figure 20. In this example, it is assumed that the intensity of the x-rays emitted at the anode A remains constant and that the x-rays passing through the aperture B cover an area of 4 square inches on reaching the recording surface C_1, which is 12 inches (D) from the anode. Then, when the recording surface is moved 12 inches farther from the anode, to C_2, so that the distance from the anode is 24 inches (2D), or twice its earlier value, the x-rays will cover 16 square inches—an area four times as great as that at C_1. It follows, therefore, that the radiation per square inch on the surface at C_2 is only one-quarter of that at the level C_1. Thus, the exposure that would be adequate at C_1 must be increased four times in order to produce at C_2 a radiograph of equal density. In practice, this can be done by increasing the time or by increasing the milliamperage.

This inverse square law can be expressed algebraically as follows:

$$\frac{I_1}{I_2} = \frac{D_2{}^2}{D_1{}^2}$$

where I_1 and I_2 are the intensities at the distances D_1 and D_2, respectively.

RADIATION ABSORPTION IN THE SPECIMEN

When x-rays or gamma rays strike an absorber (Figure 21), some of the radiation is absorbed and another portion passes through undeviated. It is the intensity variation of the undeviated radiation from area to area in the specimen that forms the useful image in a radiograph. However, not all the radiation is either completely removed from the beam or transmitted. Some is deviated within the specimen from its original direction—that is, it is scattered and is nonimage-forming. This non-image-forming scattered radiation, if not carefully controlled, will expose the film and thus tend to obscure the useful radiographic image. Scattered radiation and the means for reducing its effects are discussed in detail on pages 38 to 45. Another portion of the energy in the original beam is spent in liberating electrons from the absorber. The electrons from the specimen are unimportant radiographically; those from lead screens, discussed in detail on pages 30 to 33, are very important.

X-ray Equivalency

If industrial radiography were done with mono-energetic radiation, that is, with an x-ray beam containing but a single wavelength, and if there were no scattering, the laws of absorption of x-rays by matter could be stated mathematically with great exactness. However, since a broad band of wavelengths is used and since considerable scattered radiation reaches the film, the laws can be given only in a general way.

The x-ray absorption of a specimen depends on its thickness, on its density and, most important of all, on the atomic nature of the material. It is obvious that of two specimens of similar composition, the thicker or the more dense will absorb the more radiation, necessitating an increase in kilovoltage or exposure, or both, to produce the same photographic result. However, the atomic elements in a specimen usually exert a far greater effect upon x-ray absorption than either the thickness or the density. For example, lead is about 1.5 times as dense as ordinary steel, but at 220 kV, 0.1 inch of lead absorbs as much as 1.2 inches of steel. Brass is

TABLE IV—APPROXIMATE RADIOGRAPHIC EQUIVALENCE FACTORS

Material	X-rays								Gamma Rays			
	50 kV	100 kV	150 kV	220 kV	400 kV	1000 kV	2000 kV	4 to 25 MeV	Ir 192	Cs 137	Co 60	Radium
Magnesium	0.6	0.6	0.5	0.08								
Aluminum	1.0	1.0	0.12	0.18					0.35	0.35	0.35	0.40
2024 (aluminum) alloy	2.2	1.6	0.16	0.22					0.35	0.35	0.35	
Titanium			0.45	0.35								
Steel		12.	1.0	1.0	1.0	1.0	1.0	1.0	1.0	1.0	1.0	1.0
18-8 (steel) alloy		12.	1.0	1.0	1.0	1.0	1.0	1.0	1.0	1.0	1.0	1.0
Copper		18.	1.6	1.4	1.4			1.3	1.1	1.1	1.1	1.1
Zinc			1.4	1.3	1.3			1.2	1.1	1.0	1.0	1.0
Brass*			1.4*	1.3*	1.3*	1.2*	1.2*	1.2*	1.1*	1.1*	1.1*	1.1*
Inconel X alloy		16.	1.4	1.3	1.3	1.3	1.3	1.3	1.3	1.3	1.3	1.3
Zirconium			2.3	2.0		1.0						
Lead			14.	12.		5.0	2.5	3.0	4.0	3.2	2.3	2.0
Uranium				25.				3.9	12.6	5.6	3.4	

Aluminum is taken as the standard metal at 50kV and 100 kV and **steel** at the higher voltages and gamma rays. The thickness of another metal is multiplied by the corresponding factor to obtain the approximate equivalent thickness of the standard metal. The exposure applying to this thickness of the standard metal is used.

Example: To radiograph 0.5 inch of copper at 220 kV, multiply 0.5 inch by the factor 1.4, obtaining an equivalent thickness of 0.7 inch of steel. Therefore, give the exposure required for 0.7 inch of steel.

*Tin or lead alloyed in the brass will increase these factors.

only about 1.1 times as dense as steel, yet, at 150 kV, the same exposure is required for 0.25 inch of brass as for 0.35 inch of steel. Table IV gives approximate radiographic equivalence factors. It should be emphasized that this table is only approximate and is intended merely as a guide, since it is based on a compilation of data from many sources. In a particular instance, the exact value of the radiographic equivalence factor will depend on the quality of the x-radiation and the thickness of the specimen. It will be noted from this table that the relative absorptions of the different materials are not constant but change with kilovoltage, and that as the kilovoltage increases, the differences between all materials tend to become less. In other words, as kilovoltage is increased, the radiographic absorption of a material is less and less dependent on the atomic numbers* of its constituents.

For x-rays generated at voltages upward of 1000 kV and for materials not differing too greatly

_____

*See footnote, page 7.

in atomic number (steel and copper, for example), the radiographic absorption for a given thickness of material is roughly proportional to the density of the material. However, even at high voltages or with penetrating gamma rays, the effect of composition upon absorption cannot be ignored when dealing with materials that differ widely in atomic number. For instance, the absorption of lead for 1000 kV x-rays is about five times that of an equal thickness of steel, although its density is but 1½ times as great.

The kilovoltage governs the penetrating power of an x-ray beam and hence governs the intensity of the radiation passing through the specimen. It is not possible, however, to specify a simple relation between kilovoltage and x-ray intensity because such factors as the thickness and the kind of material radiographed, the characteristics of the x-ray generating apparatus, and whether the film is used alone or with intensifying screens exert a considerable influence on this relation. The following example illustrates this point:

Data from a given exposure chart indicate that radiographs of equal density can be made of ¼-inch steel with either of the following sets of exposure conditions:

80 kilovolts, 35 milliampere-minutes

120 kilovolts, 1.5 milliampere-minutes

Thus, in this case, a 50 percent increase in kilovoltage results in a *23-fold* increase in photographically effective x-ray intensity.

Two-inch aluminum can also be radiographed at these two kilovoltages. Equal densities will result with the following exposure relations:

80 kilovolts, 17 milliampere-minutes

120 kilovolts, 2.4 milliampere-minutes

In this case, the same increase in kilovoltage results in an increase of photographically effective x-ray intensity passing through the specimen of only *seven times*. Many other examples can be found to illustrate the extreme variability of the effect of kilovoltage on x-ray intensity.

Gamma-ray Equivalency

Essentially the same considerations apply to gamma-ray absorption, since the radiations are of similar nature. It is true that some radioactive materials used in industrial radiography emit radiation that is monoenergetic, or almost so (for example, cobalt 60 and cesium 137). However, even with these sources, scattering is dependent on the size, shape, and composition of the specimen, which prevents the laws of absorption from being stated exactly. For those gamma-ray emitters (for example, iridium 192) that give off a number of discrete gamma-ray wavelengths extending over a wide energy range, the resemblance to the absorption of x-rays is even greater.

The gamma-ray absorption of a specimen depends on its thickness, density, and composition, as does its x-ray absorption. However, the most commonly used gamma-ray sources emit fairly penetrating radiations corresponding in their properties to high-voltage x-radiation. The table of radiographic equivalence factors on page 25 shows that the absorptions of the various materials for penetrating gamma rays are similar to their absorptions for high-voltage x-rays—that is, the absorptions of materials fairly close together in atomic number are roughly proportional to their densities. As with high-voltage x-rays, this is not true of materials, such as steel and lead, that differ widely in atomic number.

RADIOGRAPHIC SCREENS

Another factor governing the photographic density of the radiograph is the intensifying action of the screens used. Intensifying screens and lead screens are fully discussed in the chapter on Radiographic Screens, pages 30 to 37. It will suffice to state here that x-rays and gamma rays cause fluorescent intensifying screens to emit light that may materially lessen the exposure necessary to produce a given density. Lead screens emit electrons under the action of x-rays and gamma rays. The photographic effect of these electrons may permit a shorter exposure than would be required without lead screens.

EXPOSURE FACTOR

The "exposure factor" is a quantity that combines milliamperage (x-rays) or source strength (gamma rays), time, and distance. Numerically the exposure factor equals

$$\frac{\text{milliamperes} \times \text{time}}{\text{distance}^2} \text{ for x-rays}$$

$$\text{and } \frac{\text{curies} \times \text{time}}{\text{distance}^2} \text{ for gamma rays}$$

Radiographic techniques are sometimes given in terms of kilovoltage and exposure factor, or radioactive isotope and exposure factor. In such a case, it is necessary merely to multiply the exposure factor by the square of the distance to be used in order to find, for example, the milliampere-minutes or the curie-hours required.

DETERMINATION OF EXPOSURE FACTORS

X-rays

The focus-film distance is easy to establish by actual measurement; the milliamperage can conveniently be determined by the milliammeter supplied with the x-ray machine; and the exposure time can be accurately controlled by a good time-switch. The tube voltage, however, is difficult and inconvenient to measure accurately. Furthermore, designs of individual machines differ widely, and may give x-ray outputs of a different quality and intensity even when operated at the same nominal values of peak kilovoltage and milliamperage.

Consequently, although specified exposure techniques can be duplicated satisfactorily in the factors of focus-film distance, milliamperage, and exposure time, one apparatus may differ materially from another in the kilovoltage setting necessary to produce the same radiographic density. Because of this, the kilovoltage setting for a given technique should be determined by trial on each x-ray generator. In the preliminary tests, published exposure charts may be followed as an approximate guide. It is customary for equipment manufacturers to calibrate x-ray machines at the factory and to furnish suitable exposure charts. For the unusual problems that arise, it is desirable to record in a logbook all the data on exposure and techniques. In this way, operators will soon build up a source of information that will make them more competent to deal with difficult situations.

For developing trial exposures, a standardized technique should always be used. If this is done, any variation in the quality of the trial radiographs may then be attributed to the exposure alone. This method obviates many of the variable factors common to radiographic work.

Since an increase of kilovoltage produces a marked increase in x-ray output and penetration (see Figure 19), it is necessary to maintain a close control of this factor in order to secure radiographs of uniform density. In many types of industrial radiography where it is desirable to maintain constant exposure conditions with regard to focus-film distance, milliamperage, and exposure time, it is common practice to vary the kilovoltage in accordance with the thickness of the material to be examined so as to secure proper density in the radiographic image. Suppose, for example, it is desired to change from radiographing 1½-inch steel to radiographing 2-inch steel. The 2-inch steel will require more than 10 times the exposure in milliampere-minutes at 170 kilovolts. However,

increasing the kilovoltage to a little more than 200 will yield a comparable radiograph with the same milliampere-minutes. Thus, kilovoltage is an important variable because economic considerations often require that exposure times be kept within fairly narrow limits. It is desirable, as a rule, to *use as low a kilovoltage as other factors will permit.* In the case of certain high-voltage x-ray machines, the technique of choosing exposure conditions may be somewhat modified. For instance, the kilovoltage may be fixed rather than adjustable at the will of the operator, leaving only milliamperage, exposure time, film type, and focus-film distance as variables.

Gamma Rays

With radioactive materials, the variable factors are more limited than with x-rays. Not only is the quality of the radiation fixed by the nature of the emitter, but also the intensity is fixed by the amount of radioactive material in the particular source. The only variables under the control of operators, and the only quantities they need to determine, are the source-film distance, film type, and the exposure time. As in the case of x-radiography, it is desirable to develop trial exposures using the gamma-ray sources under standardized conditions and to record all data on exposures and techniques.

CONTRAST

In a radiograph, the various intensities transmitted by the specimen are rendered as different densities in the image. The density differences from one area to another constitute *radiographic contrast.* Any shadow or detail within the image is visible by reason of the contrast between it and its background of surrounding structures. Within appropriate limits, the greater the contrast or density differences in the radiograph, the more definitely various details will stand out. However, if overall contrast is increased too much, there is an actual loss in visibility of detail in both the thick and the thin regions of the specimen. The thick sections will be imaged at densities too low to be useful (see Chapters 7 and 16); and the thin sections, at densities too high to be viewed on the available illuminators. This principle is fully illustrated in Figure 22, which shows two radiographs of a steel stepped wedge, one (A) exposed at a high tube voltage and the other (B) at a low voltage. It is apparent that in the middle tones the differentiation in the steps is greater in the low-voltage radiograph (B) than in the high-voltage radiograph

trast if the subject contrast is very high. *With any given specimen,* the contrast of the radiograph will depend on the kilovoltage of the x-rays or the quality of the gamma rays (Figure 22), the contrast characteristics of the film, the type of screen, the density to which the radiograph is exposed, and the processing.

RADIOGRAPHIC SENSITIVITY

In the radiography of materials of approximately uniform thickness, where the range of transmitted x-ray intensities is small, a technique producing high contrast will satisfactorily render all portions of the area of interest, and the radiographic sensitivity* will be greater than with a technique producing low contrast. If, however, the part radiographed transmits a wide range of x-ray intensities, then a technique producing lower contrast may be necessary in order to record detail in all portions of radiographic sensitivity. (See Figure 23.) The interrelations among the various factors affecting radiographic sensitivity and detail visibility are discussed on pages 64 to 67.

Figure 22—**A:** 220-kV exposure. **B:** 120-kV exposure. Radiographs of steel stepped wedge having a thickness range of ¼ to ¾ inch in ⅛-inch steps.

(A). Near the end, however, the steps shown in A are much less apparent in B.

Radiographic contrast is a result of both *subject contrast* and *film contrast.* Subject contrast is governed by the range of radiation intensities transmitted by the specimen. A flat sheet of homogeneous material of nearly uniform thickness would have very low subject contrast. Conversely, a specimen with large variations in thickness, which transmits a wide range of intensities, would have high subject contrast. Overall subject contrast could be defined as the ratio of the highest to the lowest radiation intensities falling on the film. Contrast is also affected by scattered radiation, removal of which increases subject contrast.

CHOICE OF FILM

Different films have different contrast characteristics. Thus, a film of high contrast may give a radiograph of relatively low contrast if the subject contrast is very low; conversely, a film of low contrast may give a radiograph of relatively high con-

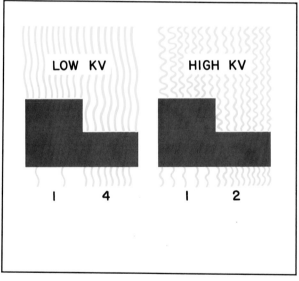

Figure 23—As kilovoltage increases, subject contrast decreases because more wavelengths penetrate the subject in both thick and thin sections, thus reducing the overall difference in exposure between the two.

---

*"Radiographic sensitivity" refers to the size of the smallest detail that can be seen in a radiograph or to the ease with which the images of small details can be detected (see Chapter 8). In comparing radiographs of the same specimen the one which renders details most clearly visible is said to have the highest radiographic sensitivity.

MULTIPLE FILM TECHNIQUES

In the radiography of specimens transmitting too wide a range of intensities to be recorded by a single high-contrast film, high radiographic contrast can be obtained by loading the cassette with *two* high-contrast films of *different speeds*. The exposure conditions are so chosen that thick portions of the specimen are satisfactorily recorded on the faster, and thin portions on the slower film.

In Figure 24, A and B, the thicker portions of the stepped wedge (½-¾ inch) are well recorded on a moderately fast film. In Figure 24B, the thinner portions (⅜-½ inch) are recorded on a film of about one-quarter the speed, exposed at the same time and in the same exposure holder. Had it been desired to render the whole of the stepped wedge in a single radiograph, the kilovoltage would have had to be increased, which would have resulted in a decrease in subject contrast, and a consequent decrease in radiographic contrast, Figure 24C.

The technique need not be limited to only two films exposed simultaneously. In special cases, three or even more films can be used, greatly expanding the range of thicknesses over which high contrast can be obtained with a single exposure.

A variation of this method is to load the cassette or film holder with two sheets of the same kind of film. With the proper exposure, details in the thick section of the specimen can be examined by viewing the two films superimposed with the images in register. The thin portions on the other hand, will be recorded on both films, either of which may be viewed alone.

Within the range of densities useful in practice, the contrasts of most industrial x-ray films increase continuously with increasing density. In the case of certain medical x-ray films that are occasionally used in industry, however, contrast first increases to a maximum with increasing density and then decreases. Thus, the contrast of a radiograph depends to a large extent upon the density. This point is discussed on pages 53 and 136 to 139.

EFFECTS OF PROCESSING

Film contrast increases with the degree of development up to a limit determined by the properties of both the film and the developer. For manual development, the development time should not be less than the minimum time recommended for the film-developer combination. Some films may be developed longer than this minimum time to obtain higher speed.

In automated processing, the recommended operating conditions of the processing machine are such that the maximum film contrast is obtained.

Radiographs of high contrast are likely to contain areas of high density in which detail cannot be seen with ordinary illumination. High-intensity illuminators, for viewing radiographs of high density, are commercially available.

A   B   C

Figure 24—A and **B:** Radiographs of stepped wedge made with a two-film technique showing thick portions (A) on a moderately fast film, and the thin portions (B) on a film about one-quarter the speed. **C:** Single radiograph at higher kilovoltage on the slower of the films. Note lower radiographic contrast in C.

CHAPTER 5

Radiographic Screens

When an x-ray or gamma-ray beam strikes a film, usually less than 1 percent of the energy is absorbed. Since the formation of the radiographic image is primarily governed by the absorbed radiation, more than 99 percent of the available energy in the beam performs no useful photographic work. Obviously, any means of more fully utilizing this wasted energy, without complicating the technical procedure, is highly desirable. Two types of radiographic screens are used to achieve this end—lead and fluorescent. Lead screens, in turn, take two different forms. One form is sheets of lead foil, usually mounted on cardboard or plastic, which are used in pairs in a conventional cassette or exposure holder. The other consists of a lead compound (usually an oxide), evenly coated on a thin support. The film is placed between the leaves of a folded sheet of this oxide-coated material with the oxide in contact with the film. The combination is supplied in a sealed, lightproof envelope.

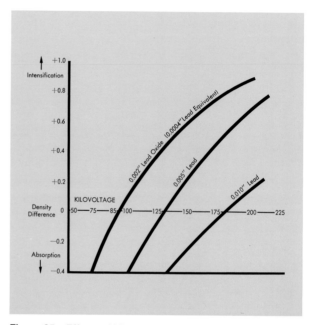

Figure 25—Effects of kilovoltage on intensification properties of lead screens.

LEAD FOIL SCREENS

For radiography in the range 150 to 400 kV, lead foil in direct contact with both sides of the film has a desirable effect on the quality of the radiograph. In radiography with gamma rays and with x-rays below 2,000 kV, the front lead foil need be only 0.004 to 0.006 inch thick; consequently its absorption of the primary beam is not serious. The back screen should be thicker to reduce backscattered radiation. Such screens are available commercially. The choice of lead screen thicknesses for multi-million-volt radiography is much more complicated, and the manufacturers of the equipment should be consulted for their recommendations.

Effects of Lead Screens

Lead foil in direct contact with the film has three principal effects: (1) It increases the photographic action on the film, largely by reason of the electrons emitted and partly by the secondary radiation generated in the lead. (2) It absorbs the longer wavelength scattered radiation more than the primary. (3) It intensifies the primary radiation more than the scattered radiation. The differential absorption of the secondary radiation and the differential intensification of the primary radiation result in diminishing the effect of scattered radiation, thereby producing greater contrast and clarity in the radiographic image. This reduction in the effect of the scattered radiation decreases the total intensity of the radiation reaching the film, thereby lessening the net intensification factor of the screens. The absorption of primary radiation by the front lead screen also diminishes the net intensifying effect, and, if the incident radiation does not have sufficient penetrating power, the actual exposure required may be even greater than without screens. At best, the exposure time is one half to one third of that without screens but the advantage of screens in reducing scattered radiation still holds.

The quality of the radiation necessary to obtain an appreciable intensification from lead foil screens depends on the type of film, the kilovoltage, and the thickness of the material through which the rays must pass. (Figure 25) In the radiography of aluminum, for example, using a 0.005-inch front screen and a 0.010-inch back screen, the thickness of aluminum must be about 6 inches and the kilovoltage as high as 160 kV to secure any advantage in exposure time with lead screens. In the radiography of steel, lead screens begin to give appreciable intensification with thicknesses in the neighborhood of ¼ inch, at voltages of 130 to 150 kV. In the radiography of 1¼ inches of steel at about 200 kV, lead screens permit an exposure of about one third of that without screens (intensification factor of 3). With cobalt 60 gamma rays, the intensification factor of lead screens is about 2. Lead foil screens, however, do not detrimentally affect the definition or graininess of the radiographic image to any material degree so long as the lead and the film are in intimate contact.

Figure 26—Upper area shows decreased density caused by paper between the lead screen and film. An electron shadow picture of the paper structure has also been introduced.

Lead foil screens diminish the effect of scattered radiation, particularly that which undercuts the object (see page 38), when the primary rays strike the portions of the film holder or cassette outside the area covered by the object.

Scattered radiation from the specimen itself is cut almost in half by lead screens, contributing to maximum clarity of detail in the radiograph; this ad-

vantage is obtained even under conditions where the lead screen makes necessary an increase in exposure. A more exhaustive discussion of scattered radiation will be found in the chapter on Scattered Radiation, which starts on page 38.

In radiography with gamma rays or high-voltage x-rays, films loaded in metal cassettes without screens are likely to record the effect of secondary electrons generated in the lead-covered back of the cassette. These electrons, passing through the felt pad on the cassette cover, produce a mottled appearance because of the structure of the felt. Films loaded in the customary lead-backed cardboard exposure holder may also show a pattern of the structure of the paper that lies between the lead and the film (Figure 26). To avoid these effects, film should be enclosed between double lead screens, care being taken to ensure good contact between film and screens. Thus, lead foil screens are essential in practically all radiography with gamma rays or million-volt x-rays. If, for any reason, screens cannot be used with these radiations, a lightproof paper or cardboard holder *with no metal backing* should be used.

Contact between the film and the lead foil screens is essential to good radiographic quality. Areas in which contact is lacking produce "fuzzy" images, as shown in Figure 27.

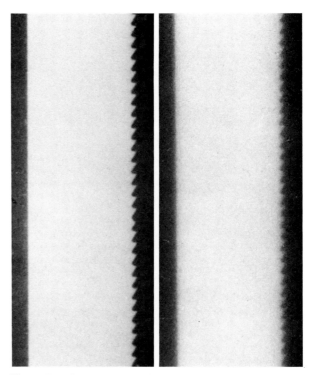

Figure 27—Good contact between film and lead foil screens gives a sharp image (left). Poor contact results in a fuzzy image (right).

31

Figure 28—Static marks resulting from poor film-handling technique. Static marks may also be treelike or branching.

Selection and Care of Lead Screens

Lead foil for screens must be selected with extreme care. Commercially pure lead is satisfactory. An alloy of 6 percent antimony and 94 percent lead, being harder and stiffer, has better resistance to wear and abrasion. Tin-coated lead foil should be avoided, since irregularities in the tin cause a variation in the intensifying factor of the screens, resulting in mottled radiographs. Minor blemishes do not affect the usefulness of the screen, but large "blisters" or cavities should be avoided.

Most of the intensifying action of a lead foil screen is caused by the electrons emitted under x-ray or gamma-ray excitation. Because electrons are readily absorbed even in thin or light materials, the surface must be kept free of grease and lint which will produce light marks on the radiograph. Small flakes of foreign material—for example, dandruff or tobacco—will likewise produce light spots on the completed radiograph. For this same reason, protective coatings on lead foil screens are not common. Any protective coating should be thin, to minimize the absorption of electrons and keep the intensification factor as high as possible, and uniform so that the intensification factor will be uniform. (In addition, the coating should not produce static electricity when rubbed against or placed in contact with film—see Figure 28.)

Deep scratches on lead foil screens, on the other hand, will result in dark lines (Figures 29 and 30).

Grease and lint may be removed from the surface of lead foil screens with a mild household detergent or cleanser and a soft, lint-free cloth. If the cleanser is one that dries to a powder, care must be taken to remove all the powder and to prevent its being introduced into the cassette or exposure holder. The screens must be completely dry before use; otherwise, the film will stick to them. If more thorough cleaning is necessary, screens may be very gently rubbed with the finest grade of steel wool. If this is done carefully, the shallow scratches left by the steel wool will not produce dark lines in the radiograph.

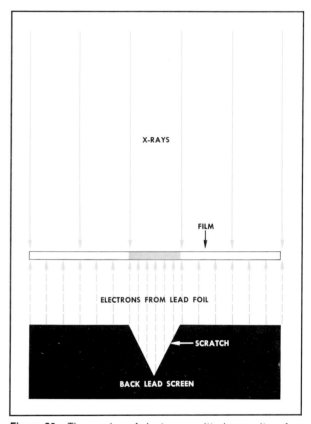

Figure 29—The number of electrons emitted per unit surface of the lead is essentially uniform. Therefore, more electrons can reach the film in the vicinity of a scratch, resulting in a dark line on the radiograph. (For illustrative clarity, electron paths have been shown as straight and parallel; actually, the electrons are emitted diffusely.)

Films may be fogged if left between lead foil screens longer than is reasonably necessary, particularly under conditions of high temperature and humidity. When screens have been freshly cleaned with an abrasive, this effect will be increased; hence, prolonged contact between film and screens should be delayed for 24 hours after cleaning.

Indications for Use of Lead Foil Screens

Lead foil screens have a wide range of usefulness. They permit a reduction in exposure time with thicknesses of metal greater than about ¼ inch of steel and kilovoltages in excess of about 130. Their use is recommended in all cases where they exhibit this intensifying action. However, they may be employed in some cases where they do not permit a reduction, but even necessitate an increase in the exposure time. The chief criterion should be the quality of the radiograph. Lead screens should be used whenever they improve radiographic quality (see Figure 31).

LEAD OXIDE SCREENS

Films packaged with lead screens in the form of lead-oxide-coated paper, factory sealed in light-tight envelopes, have a number of advantages. One of these is convenience—the time-consuming task of loading cassettes and exposure holders is avoided, as are many of the artifacts that can arise from careless handling of film.

Another advantage is cleanliness. This is particularly important in the radiography of those specimens in which heavy inclusions are serious. Light materials—hair, dandruff, tobacco ash—between the lead screen and the film will produce low-density indications on the radiograph. These can easily be confused with heavy inclusions in the specimen. The factory-sealed combination of film and lead oxide screens, manufactured under the conditions of extreme cleanliness necessary for all photographic materials, avoids all difficulties on this score and obviates much of the re-radiography that formerly was necessary. A further advantage is the flexibility of the packets, which makes them particularly valuable when film must be inserted into confined spaces.

Such packets may be used in the kilovoltage range of 100 to 300 kV. In many cases, the integral lead oxide screens will be found to have a somewhat higher intensification factor than conventional lead foil screens. They will, however, remove less scattered radiation because of the smaller effective thickness of lead in the lead oxide screen.

Scatter removal and backscatter protection equivalent to that provided by conventional lead foil screens can be provided by using conventional lead foil screens *external* to the envelope. Such screens can be protected on *both* surfaces with cardboard or plastic, and thus can be made immune to most of the accidents associated with handling. With this technique, the full function of lead foil screens can be retained while gaining the advantages of cleanliness, convenience, and screen contact.

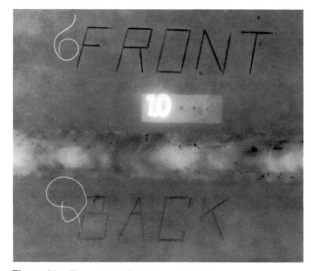

Figure 30—The words "Front" and "Back" were scratched in the surface of front and back lead foil screens before radiographing a 1-inch welded steel plate. Hairs placed between the respective screens and the film show as light marks preceding the scribed words.

Indications for Use of Lead Oxide Screens

Films packaged in this manner may be used in the kilovoltage range from 100 to 300 kV, particularly in those circumstances in which cleanliness is important and where good film-screen contact would otherwise be difficult to obtain.

FLUORESCENT SCREENS

Certain chemicals fluoresce, that is, have the ability to absorb x-rays and gamma rays and immediately emit light. The intensity of the light emitted depends on the intensity of the incident radiation. The phosphors are finely powdered, mixed with a suitable binder, and coated in a thin, smooth layer on a special cardboard or plastic support.

For the exposure, the film is clamped firmly between a pair of these screens. The photographic effect on the film, then, is the sum of the effects of the x-rays and of the light emitted by the screens. A few examples will serve to illustrate the importance of intensifying screens in reducing exposure time. In medical radiography, the exposure is from 1/10 to 1/60 as much with fluorescent intensifying screens as without them. In other words, the *intensification factor* varies from 10 to 60, depending on the kilovoltage and the type of screen used. In the radiography of ½-inch steel at 150 kV, a factor as high as 125 has been observed, and in the radiography of ¾-inch steel at 180 kV factors of several hundred have been obtained experimentally.

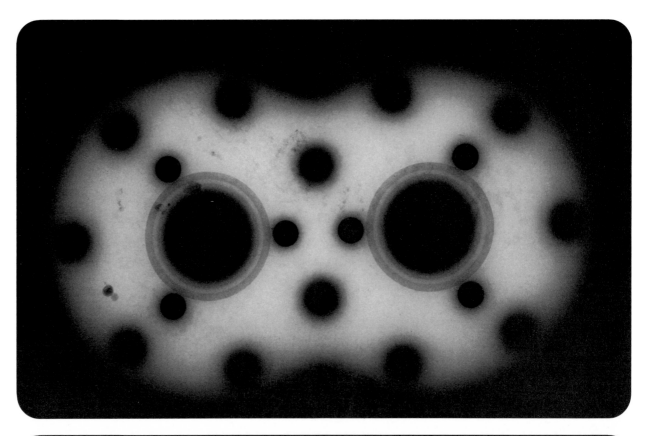

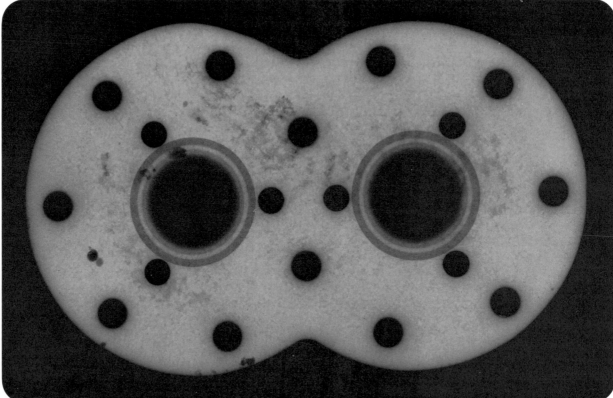

Figure 31—Lead foil screens remove scatter and increase radiographic contrast. Specimen is ¼-inch steel radiographed at 120 kV. **Above:** Without screens. **Below:** With lead foil screens. Although in this case lead foil screens necessitate some increase in exposure, they greatly improve radiographic quality.

Under these latter conditions, the intensification factor has about reached its maximum, and it diminishes both for lower voltage and thinner steel and for higher voltage and thicker steel. Using cobalt 60 gamma rays for very thick steel, the factor may be 10 or less.

Limitations

Despite their great effect in reducing exposure time, fluorescent screens are used in industrial radiography only under special circumstances. This is in part because they give poor definition in the radiograph, compared to a radiograph made directly or with lead screens. The poorer definition results from the spreading of the light emitted from the screens, as shown in Figure 32. The light from any particular portion of the screen spreads out beyond the confines of the x-ray beam that excited the fluorescence. This spreading of light from the screens accounts for the blurring of outlines in the radiograph.

The other reason fluorescent screens are used relatively little in industrial radiography is because they produce *screen mottle* on the finished radiograph. This mottle is characteristic in appearance, very much larger in scale and much "softer" in outline than the graininess associated with the film itself. It is not associated with the actual structure of the screen, that is, it is not a result of the size of the fluorescent crystals themselves or of any unevenness in their dispersion in the binder.

Rather, screen mottle is associated with purely statistical variations in the numbers of absorbed x-ray photons from one tiny area of the screen to the next. The fewer the number of x-ray photons involved, the stronger the appearance of the screen mottle. This explains, for example, why the screen mottle produced by a particular type of screen tends to become greater as the kilovoltage of the radiation increases. The higher the kilovoltage, the more energetic, on the average, are the x-ray photons. Therefore, on absorption in the screen, a larger "burst" of light is produced. The larger the bursts, the fewer that are needed to produce a given density and the greater are the purely statistical variations in the number of photons from one small area to the next. (See page 145.)

Intensifying screens may be needed in the radiography of steel thicknesses greater than about 2 inches at 250 kV, 3 inches at 400 kV, and 5 inches at 1,000 kV.

Fluorescent screens are not employed with gamma rays as a rule since, apart from the screen mottle, failure of the reciprocity law (see pages 48

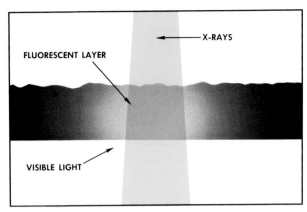

Figure 32—Diagram showing how the light and ultraviolet radiation from a typical fluorescent screen spreads beyond the x-ray beam that excites the fluorescence.

and 140) results in relatively low intensification factors with the longer exposure times usually necessary in gamma-ray radiography. In the radiography of light metals, fluorescent screens are rarely necessary; but, should they be required, the best choice would be fluorescent screens of the slowest type compatible with an economical exposure time—if possible, those designed specifically for sharpness of definition in medical radiography.

At kilovoltages higher than those necessary to radiograph about ½ inch of steel, the fastest available screens are usually employed, since the major use of fluorescent intensifying screens is to minimize the exposure time.

There are a few radiographic situations in which a speed higher than that of the fastest film designed for direct exposure or exposure with lead screens is required but which do not demand the maximum in speed given by fluorescent intensifying screens used with a film intended for that application. In such cases a high-speed, direct-exposure film may be used with fluorescent screens. The speed of this combination will be intermediate between those of the two first-mentioned combinations. However, the contrast and the maximum density will be higher than that obtained with a film designed for fluorescent-screen exposure, and the screen mottle will be less because of the lower speed of the screen-film combination.

Mounting of Fluorescent Screens

Intensifying screens are usually mounted in pairs in a rigid holder, called a cassette, so that the fluorescent surface of each screen is in direct contact with one of the emulsion surfaces of the film. Intimate contact of the screens and the film over their entire area is essential, because poor contact allows the fluorescent light to spread and produce a blurred image as shown in Figure 33, left.

As a rule, the mounting of screens is done by the x-ray dealer, who is equipped to provide this service in accordance with the manufacturer's recommendations. If the screens are mounted by the purchaser, care must be exercised to avoid physical unevenness that would result from any thick or uneven binding material. The adhesive must not cause discoloration of the screens—even a small degree of discoloration will reduce their effective speed—nor can the adhesive be such as to cause fogging of the film.

Care of Screens

Fluorescent light from intensifying screens obeys all the laws of light and cannot pass through opaque bodies as do x-rays. To prevent extraneous shadows caused by absorption of the fluorescent light by foreign matter during exposure, dust and dirt particles must not be allowed to collect between the film and screen surfaces, and stains on the screens must be avoided. Cleanliness of the order desirable for handling film and screens is sometimes difficult to maintain, but much can be done by stressing its need and eliminating carelessness.

Whenever it can be avoided, fluorescent intensifying screens should not be exposed to the full intensity of the primary beam when making a radiograph. In extreme cases, in which very high intensity primary x-radiation falls directly on the screen-film combination, the screens may become discolored or an afterglow, which will show up on subsequent radiographs, may be produced. If the specimen more than covers the screen area or if proper masking is provided, there is no difficulty from this source.

As a matter of routine, all cassettes should be tested periodically to check the contact between the screens and the film. This can be done easily by mounting a piece of wire screening (any size mesh from 1/16 inch to 1/4 inch is satisfactory) so that it lies fairly flat. The cassette is then loaded with a film, the wire screening is placed on the exposure side of the cassette, and a flash exposure is made. If there are areas of poor contact, the result will be as shown in Figure 33, left. If there is proper contact, the shadow of the wire mesh will be sharply outlined (Figure 33, right).

Close-up viewing of wire-mesh contact tests can be a very fatiguing visual task. A better and more comfortable way is for the observer to stand about 15 feet from the illuminator and look for the dark areas of the pattern, which are indicative of poor contact. Viewing is even easier and the dark areas show up more definitely if the radiograph is viewed at an angle of about 45 degrees, or if a thin sheet of light-diffusing material is put over the radiograph. Wearers of glasses often find it advantageous to remove them while viewing tests.

Fluorescent intensifying screens may be stored in the processing room but must be kept away from chemicals and other sources of contamination. The sensitive surfaces should not be touched because the images of finger marks and dust particles may show in the radiograph and interfere with accurate interpretation. Fluorescent screens usually have a transparent protective coating. This coating reduces the abrasion of the active surfaces and facilitates the removal of dirt and smudges from the screens. Every effort should be made to avoid soil-

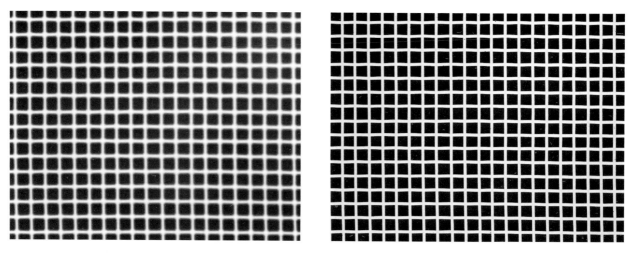

Figure 33—The sharpness of the radiographic image depends upon the contact between the intensifying screens and the film. In the radiographs of a wire mesh test object, the one (left) shows the fuzzy image produced by poor film-screen contact; the other (right), made with good contact, resulted in a sharp image.

ing fluorescent screens. Should they become dirty, they must be carefully cleaned according to the manufacturer's recommendations. Most common cleaning and bleaching agents should never be used for this purpose because their chemical composition may cause damage to the screens or fogging of the sensitive film emulsion.

The use of thin cellulose sheets for protecting the active surface of intensifying screens is particularly objectionable, because any separation between screen and film has an adverse effect on definition in the radiograph. Furthermore, under dry atmospheric conditions, merely opening the cassette is liable to produce static electrical discharges between the sheets and the film. The result will be circular or tree-like black marks in the radiograph.

Indications for Use of Fluorescent Screens

The advantage in using fluorescent intensifying screens lies in the great reductions in exposure time that their use permits. As a corollary to this, the radiography of relatively thick specimens by x-ray machines of moderate power is facilitated. For instance, using fluorescent intensifying screens, 3 inches of steel may be radiographed at 250 kV with a reasonable exposure time.

Fluorescent intensifying screens give rise to the phenomenon of screen mottle, and give poorer definition in the radiograph compared to a radiograph made directly or with lead screens. Thus, as a general rule, fluorescent screens should be used only when the exposure necessary without them would be prohibitive.

CASSETTES AND FILM HOLDERS

When intensifying or lead foil screens are used, good, uniform contact between screens and film is of prime importance. The use of vacuum cassettes is the most certain way of obtaining such intimate contact. Rigid, spring-back cassettes are also satisfactory, provided they are tested for satisfactory screen contact at reasonable intervals.

Cardboard or thin plastic exposure holders are less expensive, easier to handle in large numbers, and are flexible compared to rigid cassettes. How-

ever, if screens are to be used in them, special precautions must be taken to assure good contact. The exact means used will depend on the object radiographed. Exposure holders may be pressed or clamped against the specimen, or the weight of the specimen or the flexing of the holder as it is bent to fit some structure may provide adequate contact.

Two points should be noted, however. First, these methods do not guarantee *uniform* contact, and hence the definition of the image may vary from area to area of the film. This variation of definition may not be obvious and may cause errors in the interpretation of the radiograph. Second, such holders do not always adequately protect the film and screens from mechanical damage. A projection on the film side of the specimen may cause relatively great pressure on a small area of the film. Projections on the specimen may also give rise to pressure artifacts on the radiograph when paper-wrapped films are used. This may produce a light or dark pressure mark in the finished radiograph (Figure 34) which may be mistaken for a flaw in the specimen.

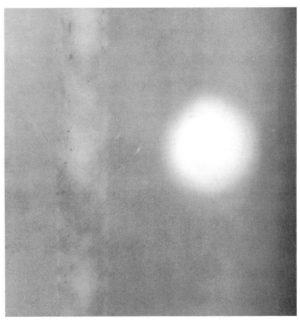

Figure 34—Low density (right) is a pressure mark, caused by a heavy object dropped on film holder before exposure.

CHAPTER 6

Scattered Radiation

When a beam of x-rays or gamma rays strikes any object, some of the radiation is absorbed, some is scattered, and some passes straight through. The electrons of the atoms constituting the object scatter radiation in all directions, much as light is dispersed by a fog. The wavelengths of much of the radiation are increased by the scattering process, and hence the scatter is always somewhat "softer," or less penetrating, than the unscattered primary radiation. Any material—whether specimen, cassette, tabletop, walls, or floor—that receives the direct radiation is a source of scattered radiation. Unless suitable measures are taken to reduce the effects of scatter, it will reduce contrast over the whole image or parts of it.

Scattering of radiation occurs, and is a problem, in radiography with both x-rays and gamma rays. In the material which follows, the discussion is in terms of x-rays, but the same general principles apply to gamma radiography.

In the radiography of thick materials, scattered radiation forms the greater percentage of the total radiation. For example, in the radiography of a ¾-inch thickness of steel, the scattered radiation from the specimen is almost twice as intense as the primary radiation; in the radiography of a 2-inch thickness of aluminum, the scattered radiation is two and a half times as great as the primary radiation. As may be expected, preventing scatter from reaching the film markedly improves the quality of the radiographic image.

As a rule, the greater portion of the scattered radiation affecting the film is from the specimen under examination (A in Figure 35). However, any portion of the film holder or cassette that extends beyond the boundaries of the specimen and thereby receives direct radiation from the x-ray tube, also becomes a source of scattered radiation which can affect the film. The influence of this scatter is most noticeable just inside the borders of the image (B in Figure 35). In a similar manner, primary radiation striking the film holder or cassette through a thin portion of the specimen will cause scattering into the shadows of the adjacent thicker portions. Such scatter is called *undercut*. Another source of scatter that may undercut a specimen is shown as C in Figure 35. If a filter is used near the tube, this too will scatter x-rays. However, because of the distance from the film, scattering from this source is of negligible importance. Any other material, such as a wall or floor, on the film side of the specimen may also scatter an

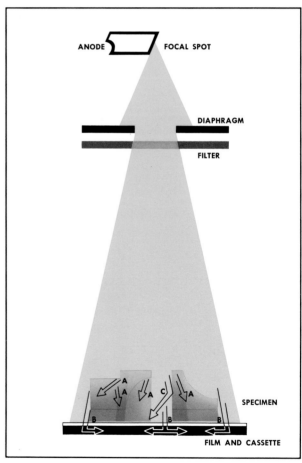

Figure 35—Sources of scattered radiation. **A:** Transmitted scatter. **B:** Scatter from cassette. **C:** "Reflection" scatter.

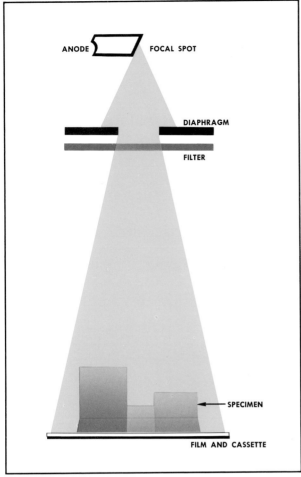

ANODE — FOCAL SPOT

DIAPHRAGM

FILTER

SPECIMEN

FILM AND CASSETTE

Figure 38—A filter placed near the x-ray tube reduces subject contrast and eliminates much of the secondary radiation, which tends to obscure detail in the periphery of the specimen.

Region	Specimen Thickness (inches)	Percentage of Original X-ray Intensity Remaining After Addition of Filter
Outside specimen	0	less than 5%
Thin section	¼	about 30%
Medium section	½	about 40%
Thick section	1	about 55%

Note that the greatest *percentage change* in x-ray intensity is under the thinner parts of the specimen and in the film area immediately surrounding it. The filter reduces by a large ratio the x-ray intensity passing through the thin sections or striking the cassette around the specimen, and hence reduces the undercut of scatter from these sources. Thus, *in*

regions of strong undercut, the contrast is *increased* by the use of a filter since the only effect of the undercutting scattered radiation is to obscure the desired image. In regions where the undercut is negligible, a filter has the effect of *decreasing* the contrast in the finished radiograph.

Although frequently the highest possible contrast is desired, there are certain instances in which too much contrast is a definite disadvantage. For example, it may be desired to render detail visible in all parts of a specimen having wide variations of thickness. If the exposure is made to give a usable density under the thin part, the thick region may be underexposed. If the exposure is adjusted to give a suitable density under the thick parts, the image of the thin sections may be grossly overexposed.

A filter reduces excessive subject contrast (and hence radiographic contrast) by hardening the radiation. The longer wavelengths do not penetrate the filter to as great an extent as do the shorter wavelengths. Therefore, the beam emerging from the filter contains a higher proportion of the more penetrating wavelengths. Figure 39 illustrates this graphically. In the sense that a more penetrating beam is produced, filtering is analogous to increasing the kilovoltage. However, it requires a comparatively large change in kilovoltage to change the hardness of an x-ray beam to the same extent as will result from adding a small amount of filtration.

Although filtering reduces the total quantity of radiation, most of the wavelengths removed are those that would not penetrate the thicker portions of the specimen in any case. The radiation removed

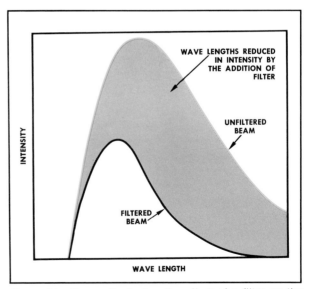

WAVE LENGTHS REDUCED IN INTENSITY BY THE ADDITION OF FILTER

UNFILTERED BEAM

INTENSITY

FILTERED BEAM

WAVE LENGTH

Figure 39—Curves illustrating the effect of a filter on the composition and intensity of an x-ray beam.

would only result in a high intensity in the regions around the specimen and under its thinner sections, with the attendant scattering undercut and overexposure. The harder radiation obtained by filtering the x-ray beam produces a radiograph of lower contrast, thus permitting a wider range of specimen thicknesses to be recorded on a single film than would otherwise be possible.

Thus, a filter can act either to increase or to decrease the net contrast. The contrast and penetrameter visibility (see page 67) are *increased* by the removal of the scatter that undercuts the specimen (see Figure 40) and *decreased* by the hardening of the original beam. The nature of the individual specimen will determine which of these effects will predominate or whether both will occur in different parts of the same specimen.

The choice of a filter material should be made on the basis of availability and ease of handling. For the same filtering effect, the thickness of filter required is less for those materials having higher absorption. In many cases, copper or brass is the most useful, since filters of these materials will be thin enough to handle easily, yet not so thin as to be delicate. See Figure 41.

Definite rules as to filter thicknesses are difficult to formulate exactly because the amount of filtration required depends not only on the material and thickness range of the specimen, but also on the distribution of material in the specimen and on the amount of scatter undercut that it is desired to eliminate. In the radiography of aluminum, a filter of copper about 4 percent of the greatest thickness of the specimen should prove the thickest necessary. With steel, a copper filter should ordinarily be about 20 percent, or a lead filter about 3 percent, of the greatest specimen thickness for the greatest useful filtration. The foregoing values are maxi-

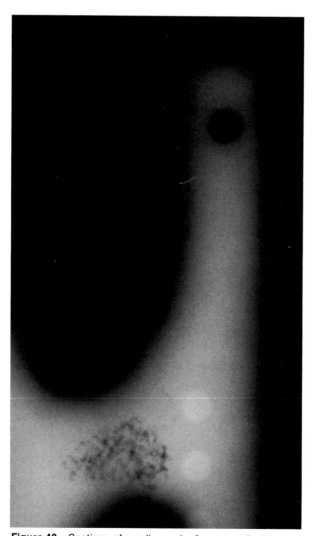 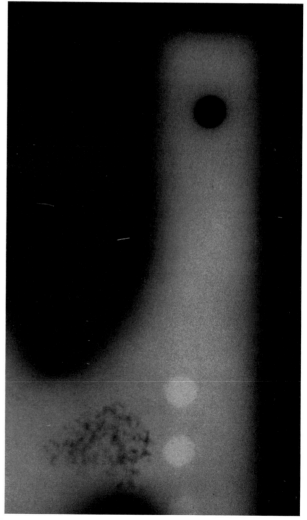

Figure 40—Sections of a radiograph of an unmasked 1⅛-inch casting, made at 200 kV without filtration (left), and as improved by filtration at the tube (right).

mum values, and, depending on circumstances, useful radiographs can often be made with far less filtration.

In radiography with x-rays up to at least 250 kV, the 0.005-inch front lead screen customarily used is an effective filter for the scatter from the bulk of the specimen. Additional filtration between specimen and film only tends to contribute additional scatter from the filter itself. The scatter undercut can be decreased by adding an appropriate filter at the tube as mentioned before (see also Figure 40). Although the filter near the tube gives rise to scattered radiation, the scatter is emitted in all directions, and since the film is far from the filter, scatter reaching the film is of very low intensity.

Further advantages of placing the filter near the x-ray tube are that specimen-film distance is kept to a minimum and that scratches and dents in the filter are so blurred that their images are not apparent on the radiograph.

Grid Diaphragms

One of the most effective ways to reduce scattered radiation from an object being radiographed is through the use of a Potter-Bucky diaphragm. This apparatus (Figure 42, page 44) consists of a moving grid, composed of a series of lead strips held in position by intervening strips of a material transparent to x-rays. The lead strips are tilted, so that the plane of each is in line with the focal spot of the tube. The slots between the lead strips are several times as deep as they are wide. The parallel lead strips have the function of absorbing the very divergent scattered rays from the object being radiographed, so that most of the exposure is made by the primary rays emanating from the focal spot of the tube and passing between the lead strips. During the course of the exposure, the grid is moved, or oscillated, in a plane parallel to the film as shown by the black arrows in Figure 42. Thus, the shadows of the lead strips are blurred out so that they do not appear in the final radiograph.

The use of the Potter-Bucky diaphragm in industrial radiography complicates the technique to some extent and necessarily limits the flexibility of the arrangement of the x-ray tube, the specimen, and the film. Grids can, however, be of great value in the radiography of beryllium more than about 3 inches thick and in the examination of other low-absorption materials of moderate and great thicknesses. For these materials, kilovoltages in the medical radiographic range are used, and the medical forms of Potter-Bucky diaphragms are appropriate. Grid ratios (the ratio of height to width of

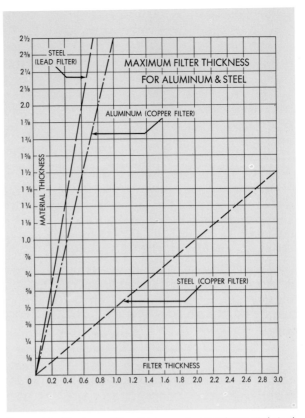

Figure 41—Maximum filter thickness for aluminum and steel.

the *openings* between the lead strips) of 12 or more are desirable.

The Potter-Bucky diaphragm is seldom used elsewhere in the industrial field, although special forms have been designed for the radiography of steel with voltages as high as 200 to 400 kV. These diaphragms are not used at higher voltages or with gamma rays because relatively thick lead strips would be needed to absorb the radiation scattered at these energies. This in turn would require a Potter-Bucky diaphragm, and the associated mechanism, of an uneconomical size and complexity.

MOTTLING CAUSED BY X-RAY DIFFRACTION

A special form of scattering caused by x-ray diffraction (see pages 129 to 131) is encountered occasionally. It is most often observed in the radiography of fairly thin metallic specimens whose grain size is large enough to be an appreciable fraction of the part thickness. The radiographic appearance of this type of scattering is mottled and may be confused with the mottled appearance sometimes produced by porosity or segregation. It can be distinguished from these conditions by making two successive radiographs, with the specimen rotated slightly (1 to 5 degrees) between exposures, about an axis perpendicular to the central beam. A pattern caused by porosity or segregation will change only slightly; however, one caused by diffraction will show a marked change. The radiographs of some specimens will show a mottling from both effects, and careful observation is needed to differentiate between them.

The basic facts of x-ray diffraction are given on pages 129 to 131. Briefly, however, a relatively large crystal or grain in a relatively thin specimen may in some cases "reflect" an appreciable portion of the x-ray energy falling on the specimen, much as if it were a small mirror. This will result in a light spot on the developed radiograph corresponding to the position of the particular crystal and may also produce a dark spot in another location if the diffracted, or "reflected," beam strikes the film. Should this beam strike the film beneath a thick part of the specimen, the dark spot may be mistaken for a void in the thick section. This effect is not observed in most industrial radiography, because most specimens are composed of a multitude of very minute crystals or grains, variously oriented; hence, scatter by diffraction is essentially uniform over the film area. In addition, the directly transmitted beam usually reduces the contrast in the diffraction pattern to a point where it is no longer visible on the radiograph.

The mottling caused by diffraction can be reduced, and in some cases eliminated, by raising the kilovoltage and by using lead foil screens. The former is often of positive value even though the radiographic contrast is reduced. Since definite rules are difficult to formulate, both approaches should be tried in a new situation, or perhaps both used together.

It should be noted, however, that in some instances, the presence or absence of mottling caused by diffraction has been used as a rough indication of grain size and thus as a basis for the acceptance or the rejection of parts.

SCATTERING IN 1- AND 2-MILLION-VOLT RADIOGRAPHY

Lead screens should always be used in this voltage range. The common thicknesses, 0.005-inch front and 0.010-inch back, are both satisfactory and convenient. Some users, however, find a 0.010-inch front screen of value because of its greater selective absorption of the scattered radiation from the specimen.

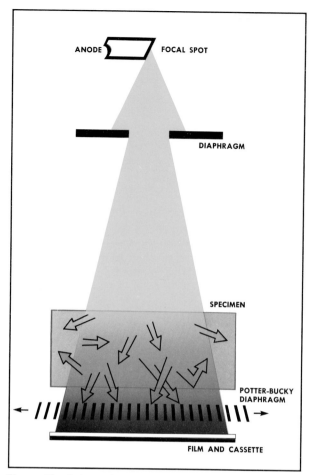

Figure 42—Schematic diagram showing how the primary x-rays pass between the lead strips of the Potter-Bucky diaphragm while most of the scattered x-rays are absorbed because they strike the sides of the strips.

Filtration at the tube offers no improvement in radiographic quality. However, filters at the film improve the radiograph in the examination of uniform sections, but give poor quality at the edges of the image of a specimen because of the undercut of scattered radiation from the filter itself. Hence, filtration should not be used in the radiography of specimens containing narrow bars, for example, no matter what the thickness of the bars in the direction of the primary radiation. Further, filtration should be used only where the film can be adequately protected against backscattered radiation.

Lead filters are most convenient for this voltage range. When thus used between specimen and film, filters are subject to mechanical damage. Care should be taken to reduce this to a minimum, lest filter defects be confused with structures in or on the specimen. In radiography with million-volt x-rays, specimens of uniform sections may be conveniently divided into three classes. Below about

1½ inches of steel, filtration affords little improvement in radiographic quality. Between 1½ and 4 inches of steel, the thickest filter, up to ⅛-inch lead, that at the same time allows a reasonable exposure time, may be used. Above 4 inches of steel, filter thicknesses may be increased to ¼ inch of lead, economic considerations permitting. It should be noted that in the radiography of extremely thick specimens with million-volt x-rays, fluorescent screens (see pages 33 to 35) may be used to increase the photographic speed to a point where filters can be used without requiring excessive exposure time.

A very important point is to block off all radiation except the useful beam with heavy (½-inch to 1-inch) lead at the anode. Unless this is done, radiation striking the walls of the x-ray room will scatter back in such quantity as to seriously affect the quality of the radiograph. This will be especially noticeable if the specimen is thick or has parts projecting relatively far from the film.

MULTIMILLION-VOLT RADIOGRAPHY

Techniques of radiography in the 6- to 24-million-volt range are difficult to specify. This is in part because of the wide range of subjects radiographed, from thick steel to several feet of mixtures of solid organic compounds, and in part because the sheer size of the specimens and the difficulty in handling them often impose limitations on the radiographic techniques that can be used.

In general, the speed of the film-screen combination increases with increasing thickness of front and back lead screens up to at least 0.030 inch. One problem encountered with screens of such great thickness is that of screen contact. For example, if a conventional cardboard exposure holder is supported vertically, one or both of the heavy screens may tend to sag away from the film, with a resulting degradation of the image quality. Vacuum cassettes are especially useful in this application and several devices have been constructed for the purpose, some of which incorporate such refinements as automatic preprogrammed positioning of the film behind the various areas of a large specimen.

The electrons liberated in lead by the absorption of multimegavolt x-radiation are very energetic. This means that those arising from fairly deep within a lead screen can penetrate the lead, being scattered as they go, and reach the film. Thus, when thick screens are used, the electrons reaching the film are "diffused," with a resultant deleterious effect on image quality. Therefore, when the highest quality is required in multimillion-volt radiography, a comparatively thin front screen (about 0.005 inch) is used, and the back screen is eliminated. This necessitates a considerable increase in exposure time. Naturally, the applicability of the technique depends also on the amount of backscattered radiation involved and is probably not applicable where large amounts occur.

Arithmetic of Exposure

CHAPTER 7

RELATIONS OF MILLIAMPERAGE (SOURCE STRENGTH), DISTANCE, AND TIME

It was pointed out on pages 22 to 24 that with a given kilovoltage of x-radiation or with the gamma radiation from a particular isotope, the three factors governing the exposure are the milliamperage (for x-rays) or source strength (for gamma rays), time, and source-film distance. The numerical relations among these three quantities are demonstrated below, using x-rays as an example. The same relations apply for gamma rays, provided the number of curies in the source is substituted wherever milliamperage appears in an equation.

The necessary calculations for any changes in focus-film distance (D), milliamperage (M), or time (T) are matters of simple arithmetic and are illustrated in the following example. As noted earlier, kilovoltage changes cannot be calculated directly but must be obtained from the exposure chart of the equipment or the operator's logbook.

All of the equations shown on these pages can be solved easily for any of the variables (mA, T, D), using one basic rule of mathematics: If one factor is moved across the equals sign (=), it moves from the numerator to the denominator or vice versa.

Example: $\dfrac{A}{B} \diagdown\!\!\!\!\diagup \dfrac{C}{D}$

To solve for B:

①$\dfrac{A}{\boxed{B}} \diagup C \diagdown \dfrac{}{D}$ ②$A \diagdown \dfrac{BC}{\boxed{D}}$ ③$AD \diagup \dfrac{}{} B\!\!\bigcirc\!\!C$ ④$\dfrac{AD}{C} = B$

*$\bigcirc$ = move this value next in the direction indicated by the arrow.

We can now solve for any unknown by:

1. Eliminating any factor that remains constant (has the same value and is in the same location on both sides of the equation).

2. Simplifying the equation by moving the unknown value so that it is alone on one side of the equation in the numerator.

3. Substituting the known values and solving the equation.

Milliamperage-Distance Relation

The milliamperage employed in any exposure technique should be in conformity with the manufacturer's rating of the x-ray tube. In most laboratories, however, a constant value of milliamperage is usually adopted for convenience.

Rule: *The milliamperage (M) required for a given exposure is directly proportional to the square of the focus-film distance (D).* The equation is expressed as follows:

$$M_1 : M_2 :: D_1{}^2 : D_2{}^2 \text{ or } \frac{M_1}{M_2} = \frac{D_1{}^2}{D_2{}^2}$$

Example: Suppose that with a given exposure time and kilovoltage, a properly exposed radiograph is obtained with 5mA (M_1) at a distance of 12 inches (D_1), and that it is desired to increase the sharpness of detail in the image by increasing the focus-film distance to 24 inches (D_2). The correct milliamperage (M_2) to obtain the desired radiographic density at the increased distance (D_2) may be computed from the proportion:

$$5 : M_2 :: 12^2 : 24^2 \text{ or } \frac{5}{M_2} = \frac{12^2}{24^2}$$

$$M_2 = 5 \times \frac{24^2}{12^2}$$

$$M_2 = 20 \text{ mA}$$

When very low kilovoltages, say 20 kV or less, are used, the x-ray intensity decreases with distance more rapidly than calculations based on the inverse square law would indicate because of absorption of the x-rays by the air. Most industrial radiography, however, is done with radiation so penetrating that the air absorption need not be considered. These comments also apply to the time-distance relations discussed below.

Time-Distance Relation

Rule: *The exposure time (T) required for a given exposure is directly proportional to the square of the focus-film distance (D).* Thus:

$$T_1 : T_2 :: D_1{}^2 : D_2{}^2 \text{ or } \frac{T_1}{T_2} = \frac{D_1{}^2}{D_2{}^2}$$

To solve for either a new Time (T_2) or a new Distance (D_2), simply follow the steps shown in the example on page 46.

Tabular Solution of Milliamperage-Time and Distance Problems

Problems of the types discussed above may also be solved by the use of a table similar to Table V. The factor between the new and the old exposure time, milliamperage, or milliampere-minute (mA-min) value appears in the box at the intersection of the column for the new source-film distance and the row for the old source-film distance.

Suppose, for example, a properly exposed radiograph is produced with an exposure of 20 mA-min with a source-film distance of 30 inches. It is desired to increase the source-film distance to 45 inches in order to decrease the geometric unsharpness in the radiograph. The factor appearing in the box at the intersection of the column for 45″ (new

TABLE V—MILLIAMPERAGE-TIME AND DISTANCE RELATIONS

Old Dist. \ New Dist.	25″	30″	35″	40″	45″	50″	55″	60″	65″	70″	75″	80″
25″	1.0	1.4	2.0	2.6	3.2	4.0	4.8	5.6	6.8	7.8	9.0	10.0
30″	0.70	1.0	1.4	1.8	2.3	2.8	3.4	4.0	4.8	5.4	6.3	7.1
35″	0.51	0.74	1.0	1.3	1.6	2.0	2.5	3.0	3.4	4.0	4.6	5.2
40″	0.39	0.56	0.77	1.0	1.3	1.6	1.9	2.2	2.6	3.1	3.5	4.0
45″	0.31	0.45	0.60	0.79	1.0	1.2	1.5	1.8	2.1	2.4	2.8	3.2
50″	0.25	0.36	0.49	0.64	0.81	1.0	1.2	1.4	1.7	2.0	2.2	2.6
55″	0.21	0.30	0.40	0.53	0.67	0.83	1.0	1.2	1.4	1.6	1.9	2.1
60″	0.17	0.25	0.34	0.44	0.56	0.69	0.84	1.0	1.2	1.4	1.6	1.8
65″	0.15	0.21	0.29	0.38	0.48	0.59	0.72	0.85	1.0	1.2	1.3	1.5
70″	0.13	0.18	0.25	0.33	0.41	0.51	0.62	0.74	0.86	1.0	1.1	1.3
75″	0.11	0.16	0.22	0.28	0.36	0.45	0.54	0.64	0.75	0.87	1.0	1.1
80″	0.10	0.14	0.19	0.25	0.32	0.39	0.47	0.56	0.66	0.77	0.88	1.0

source-film distance) and the row for 30" (old source-film distance) is 2.3. Therefore, the old milliampere-minute value (20) should be multiplied by 2.3 to give the new value—46 mA-min.

Note that some approximation is involved in the use of such a table, since the values in the boxes are rounded off to two significant figures. However, the errors involved are always less than 5 percent and, in general, are insignificant in actual practice.

Further, a table of the type of Table V obviously cannot include all source-film distances, because of limitations of space. However, in any one radiographic department, only a few source-film distances are used in the great bulk of the work, and a table of reasonable size can be constructed involving only these few distances.

Milliamperage-Time Relation

Rule: *The milliamperage (M) required for a given exposure is inversely proportional to the time (T):*

$$M_1 : M_2 : : T_2 : T_1 \text{ or } \frac{M_1}{M_2} = \frac{T_2}{T_1}$$

Another way of expressing this is to say that for a given set of conditions (voltage, distance, etc), the product of milliamperage and time is constant for the same photographic effect.

Thus, $M_1T_1 = M_2T_2 = M_3T_3 = C$, a constant.

This is commonly referred to as the *reciprocity law*. (Important exceptions are discussed below.)

To solve for either a new time (T_2) or a new milliamperage (M_2), simply follow the steps shown in the example on page 46.

THE RECIPROCITY LAW

In the sections immediately preceding, it has been assumed that exact compensation for a decrease in the time of exposure can be made by increasing the milliamperage according to the relation $M_1T_1 = M_2T_2$. This may be written MT = C and is an example of the general photochemical law that the same effect is produced for IT = constant, where I is intensity of the radiation and T is the time of exposure. It is called the *reciprocity law* and is *true for direct x-ray and lead screen exposures.* For exposures to light, it is not quite accurate and, since some radiographic exposures are made with the light from fluorescent intensifying screens, the law cannot be strictly applied.

Errors as the result of assuming the validity of the reciprocity law are usually so small that they are not noticeable in examples of the types given in the preceding sections. Departures may be apparent, however, if the intensity is changed by a factor of 4 or more. Since intensity may be changed by changing the source-film distance, failure of the reciprocity law may appear to be a violation of the inverse square law. Applications of the reciprocity law over a wide intensity range sometimes arise, and the relation between results and calculations may be misleading unless the possibility of failure of the reciprocity law is kept in mind. Failure of the reciprocity law means that the *efficiency* of a light-sensitive emulsion in utilizing the light energy depends on the light *intensity. Under the usual conditions of industrial radiography, the number of milliampere-minutes required for a properly exposed radiograph made with fluorescent intensifying screens increases as the x-ray intensity decreases, because of reciprocity failure.*

If the milliamperage remains constant and the x-ray intensity is varied by changing the focus-film distance, the compensating changes shown in Table VI should be made in the exposure time.

Table VI gives a rough estimate of the deviations from the rules given in the foregoing section that are necessitated by failure of the reciprocity law for exposures with fluorescent intensifying screens. It must be emphasized that the figures in column 3 are only approximate. The exact values of the factors vary widely with the intensity of the fluorescent light and with the density of the radiograph.

When distance is held constant, the milliamperage may be increased or decreased by a factor of 2, and the new exposure time may be calculated by the method shown on page 47, without introducing errors caused by failure of the reciprocity law that are serious in practice.

LOGARITHMS

Since logarithms are used a great deal in the following section, a brief discussion of them is included here. A more detailed treatment will be found in some handbooks and intermediate algebra texts.

Before discussing logarithms, it is necessary to define the term "power." The power of a number is the product obtained when the number is multiplied by itself a given number of times. Thus $10^3 = 10 \times 10 \times 10 = 1000$; $5^2 = 5 \times 5 = 25$. In the first example, 1000 is the third power of 10; in the second, 25 is the second power of 5, or 5 raised to the second power. The superscript figure ^{2} is known as the *exponent.* Fractional exponents are used to denote roots.

For example: $n^{\frac{a}{b}} = \sqrt[b]{n^a}$; $16^{\frac{1}{2}} = \sqrt{16} = 4$

$$10^{2.50} = 10^{\frac{5}{2}} = \sqrt[2]{10^5} = 316$$

Negative exponents indicate reciprocals of powers.

Thus: $n^{-a} = \dfrac{1}{n^a}$; $10^{-2} = \dfrac{1}{10^2} = \dfrac{1}{100} = 0.01$

The *common logarithm* of a number is the exponent of the power to which 10 must be raised to give the number in question. For example, the logarithm of 100 is 2. The logarithm of 316 equals 2.50, or log 316 = 2.50; the logarithm of 1000 equals 3, or log 1000 = 3. It is also said that 1000 is the antilogarithm of 3 or antilog 3 = 1000.

Logarithms consist of two parts: A decimal which is always positive, called the *mantissa;* and an integer which may be positive or negative, called the *characteristic*. In the case of log 316 = 2.50, .50 is the mantissa and 2 is the characteristic. The mantissa may be found by reference to a table of logarithms, by the use of a slide rule (D and L scales), or by reference to Figure 43. Regardless of the location of the decimal point, the logarithms of all numbers having the same figures in the same order will have the same mantissa.

The characteristic of the logarithm is determined by the location of the decimal point in the number. If the number is greater than one, the characteristic is positive and its numerical value is one *less* than the number of digits to the left of the decimal point. If the number is less than one (for example, a decimal fraction), the characteristic is negative and has a numerical value of one *greater* than the number of zeros between the decimal point and the first integer. A negative characteristic of 3 is written either as $\bar{3} \cdots$ to indicate that only the characteristic is negative, or as $7 \cdots -10$.

From Figure 43, we see that the mantissa of the logarithm of 20 is 0.30. The characteristic is 1.

TABLE VI—APPROXIMATE CORRECTIONS FOR RECIPROCITY LAW FAILURE

Distance Increased by	(Direct and Lead Screen Exposures) Exposure Time Multiplied by*	(Fluorescent Screen Exposures) Exposure Time Multiplied by*
25%	1.6	about 2
50	2.3	about 4
100	4.0	about 8
Distance Decreased by		
20%	0.62	about 0.5
33	0.43	about 0.2
50	0.25	about 0.1

*Column 2 shows the changes necessitated by the inverse square law only. Column 3 shows the combined effects of the inverse square law and failure of the reciprocity law.

log 20	= 1.30
log 40	= 1.60
log 80	= 1.90
log 160	= 2.20
log 200	= 2.30
log 2000	= 3.30
log 20000	= 4.30
log 0.2	= $\bar{1}.30$ or $9.30 - 10$
log 0.02	= $\bar{2}.30$ or $8.30 - 10$

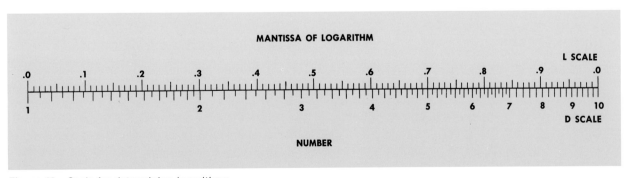

Figure 43—Scale for determining logarithms.

The preceding tabulation illustrates a very important property of logarithms. Note that when a series of numbers increases by a constant *factor*, for example, the series 20, 40, 80, 160 or the series 20, 200, 2000, 20,000, the logarithms have a constant *difference*, in these cases 0.30 and 1.00, respectively. In other words, a constant *increase* in the logarithm of a number means a constant *percentage increase* in the number itself.

PHOTOGRAPHIC DENSITY

Photographic density refers to the quantitative measure of film blackening. When no danger of confusion exists, photographic density is usually spoken of merely as *density*. Density is defined by the equation

$$D = \log \frac{I_o}{I_t}$$

where D is density, I_o is the light intensity incident on the film and I_t is the light intensity transmitted.

The tabulation below illustrates some relations between transmittance, percent transmittance, opacity, and density.

Trans- mittance $\left(\dfrac{I_t}{I_o}\right)$	Percent Trans- mittance $\left(\dfrac{I_t}{I_o} \times 100\right)$	Opacity $\left(\dfrac{I_o}{I_t}\right)$	Density $\left(\text{Log } \dfrac{I_o}{I_t}\right)$
1.00	100	1	0
0.50	50	2	0.3
0.25	25	4	0.6
0.10	10	10	1.0
0.01	1	100	2.0
0.001	0.1	1,000	3.0
0.0001	0.01	10,000	4.0

This table shows that an increase in density of 0.3 reduces the light transmitted to one-half its former value. In general, since density is a logarithm, a certain *increase* in density always corresponds to the same *percentage decrease* in transmittance.

DENSITOMETERS

A densitometer is an instrument for measuring photographic densities. A number of different types, both visual and photoelectric, are available commercially. For purposes of practical industrial radiography there is no great premium on high accuracy of a densitometer. A much more important property is reliability, that is, the densitometer should reproduce readings from day to day.

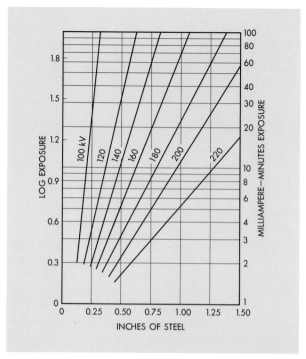

Figure 44—Typical exposure chart for steel. This chart may be taken to apply to Film X (Figure 47, for example), with lead foil screens, at a film density of 1.5. Source-film distance, 40 inches.

X-RAY EXPOSURE CHARTS

An exposure chart is a graph showing the relation between material thickness, kilovoltage, and exposure. In its most common form, an exposure chart resembles Figure 44. These graphs are adequate for determining exposures in the radiography of uniform plates, but they serve only as rough guides for objects, such as complicated castings, having wide variations of thickness.

Exposure charts are usually available from manufacturers of x-ray equipment. Because, in general, such charts cannot be used for different x-ray machines unless suitable correction factors are applied, individual laboratories sometimes prepare their own.

PREPARING AN EXPOSURE CHART

A simple method for preparing an exposure chart is to make a series of radiographs of a pile of plates consisting of a number of steps. This "step tablet," or stepped wedge, is radiographed at several different exposure times at each of a number of kilovoltages. The exposed films are all processed under conditions identical to those that will later be used for routine work. Each radiograph consists of a series of photographic densities corresponding to the x-ray intensities transmitted by the different thicknesses of metal. A certain density, for example, 1.5, is selected as the basis for the preparation of the chart. Wherever this density occurs on the stepped-wedge radiographs, there are corresponding values of thickness, milliampere-minutes, and kilovoltage. It is unlikely that many of the radiographs will contain a value of exactly 1.5 in density, but the correct thickness for this density can be found by interpolation between steps. Thickness and milliampere-minute values are plotted for the different kilovoltages in the manner shown in Figure 44.

Another method, requiring fewer stepped-wedge exposures but more arithmetical manipulation, is to make one step-tablet exposure at each kilovoltage and to measure the densities in the processed stepped-wedge radiographs. The exposure that would have given the chosen density (in this case 1.5) under any particular thickness of the stepped wedge can then be determined from the characteristic curve of the film used (see pages 53 and 54). The values for thickness, kilovoltage, and exposure are plotted as described on page 50.

Note that thickness is on a linear scale, and that milliampere-minutes are on a logarithmic scale. The logarithmic scale is not necessary, but it is very convenient because it compresses an otherwise long scale. A further advantage of the logarithmic exposure scale is that it usually allows the location of the points for any one kilovoltage to be well approximated by a straight line.

Any given exposure chart applies to a set of specific conditions. These fixed conditions are:

1. The x-ray machine used

2. A certain source-film distance

3. A particular film type

4. Processing conditions used

5. The film density on which the chart is based

6. The type of screens (if any) that are used

Only if the conditions used in making the radiograph agree in all particulars with those used in preparation of the exposure chart can values of exposure be read directly from the chart. Any change requires the application of a correction factor. The correction factor applying to each of the conditions listed previously will be discussed separately:

1. It is sometimes difficult to find a correction factor to make an exposure chart prepared for one x-ray machine applicable to another. Different x-ray machines operating at the same nominal kilovoltage and milliamperage settings may give not only different intensities but also different qualities of radiation.

2. A change in source-film distance may be compensated for by the use of the inverse square law (page 23) or, if fluorescent screens are used, by reference to the tabulation on page 49. Some exposure charts give exposures in terms of "exposure factor" (see page 26) rather than in terms of milliampere-minutes or milliampere-seconds. Charts of this type are readily applied to any value of source-film distance.

3. The use of a different type of film can be corrected for by comparing the difference in the amount of exposure necessary to give the same density on both films from relative exposure charts such as those shown in Figure 47.

   For example, to obtain a density of 1.5 using Film Y, 0.6 more exposure is required than for Film X.

   This log exposure difference is found on the L scale* and corresponds to an exposure factor of 3.99 on the D scale. (Read directly below the log E difference.) Therefore, in order to obtain the same density on Film Y as on Film X, multiply the original exposure by 3.99 to get the new exposure. Conversely, if going from Film Y to Film X, divide the original exposure by 3.99 to obtain the new exposure.

   These procedures can be used to change densities on a single film as well. Simply find the log E difference needed to obtain the new density on the film curve; read the corresponding exposure factor from the chart; then multiply to increase density or divide to decrease density.

4. A change in processing conditions causes a change in effective film speed. If the processing of the radiographs differs from that used for the exposures from which the chart was made, the correction factor must be found by experiment.

5. The chart gives exposures to produce a certain density. If a different density is required, the correction factor may be calculated from the film's characteristic curve (see pages 52 to 57).

6. If the type of screens is changed, for example from lead foil to fluorescent, it is easier and more accurate to make a new exposure chart than to attempt to determine correction factors.

---

*See Figure 43 for the logarithmic scales used in this example.

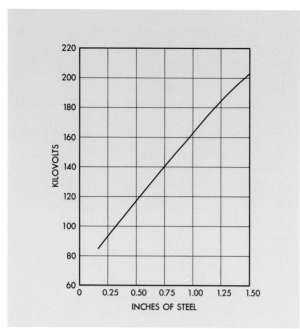

Figure 45—Typical exposure chart for use when exposure and distance are held constant and kilovoltage is varied to conform to specimen thickness. Film X (Figure 47, for example), exposed with lead foil screens to a density of 1.5. Source-film distance, 40 inches; exposure, 50 mA-min.

Sliding scales can be applied to exposure charts to allow for changes in one or more of the conditions discussed, with the exception of the first and the last. The methods of preparing and using such scales are described in detail on pages 57 to 60.

In some radiographic operations, the exposure time and the source-film distance are set by economic considerations or on the basis of previous experience and test radiographs. The tube current is, of course, limited by the design of the tube. This leaves as variables only the thickness of the specimen and the kilovoltage. When these conditions exist, the exposure chart may take a simplified form as shown in Figure 45, which allows the kilovoltage for any particular specimen thickness to be chosen readily. Such a chart will probably be particularly useful when uniform sections must be radiographed in large numbers by relatively untrained persons. This type of exposure chart may be derived from a chart similar to Figure 44 by following the horizontal line corresponding to the chosen milliampere-minute value and noting the thickness corresponding to this exposure for each kilovoltage. These thicknesses are then plotted against kilovoltage.

GAMMA-RAY EXPOSURE CHARTS

A typical gamma-ray exposure chart is shown in Figure 46. It is somewhat similar to Figure 44. However, with gamma rays, there is no variable factor corresponding to the kilovoltage. Therefore, a gamma-ray exposure chart contains one line, or several parallel lines, each of which corresponds to a particular film type, film density, or source-film distance. Gamma-ray exposure guides are also available in the form of linear or circular slide rules. These contain scales on which can be set the various factors of specimen thickness, source strength, and source-film distance, and from which exposure time can be read directly.

Sliding scales can also be applied to gamma-ray exposure charts of the type of Figure 46 to simplify some exposure determinations. For the preparation and use of such scales see pages 57 to 60.

THE CHARACTERISTIC CURVE*

The characteristic curve, sometimes referred to as the sensitometric curve or the H and D curve (after Hurter and Driffield who, in 1890, first used it), expresses the relation between the exposure applied to a photographic material and the resulting photographic density. The characteristic curves of three typical films, exposed between lead foil screens to x-rays, are given in Figure 47. Such curves are obtained by giving a film a series of known exposures, determining the densities produced by these exposures, and then plotting density against the logarithm of relative exposure.

Relative exposure is used because there are no convenient units, suitable to all kilovoltages and scattering conditions, in which to express radiographic exposures. Hence, the exposures given a film are expressed in terms of some particular exposure, giving a relative scale. In practical radiography, this lack of units for x-ray intensity or quantity is no hindrance, as will be seen below. The use of the logarithm of the relative exposure, rather than the relative exposure itself, has a number of advantages. It compresses an otherwise long scale. Furthermore, in radiography, ratios of exposures or intensities are usually more significant than the exposures or the intensities themselves. Pairs of exposures having the same ratio will be separated by the same interval on the log relative exposure scale, no matter what their absolute value may be. Consider the pairs of exposures following.

―――――――

*See also pages 136 to 140.

52

Relative Exposure	Log Relative Exposure	Interval in Log Relative Exposure
1	0.0	
5	0.70	} 0.70
2	0.30	
10	1.00	} 0.70
30	1.48	
150	2.18	} 0.70

This illustrates another useful property of the logarithmic scale. Figure 43 shows that the antilogarithm of 0.70 is 5, which is the ratio of each pair of exposures. Hence, to find the ratio of *any* pair of exposures, it is necessary only to find the antilog of the log E (logarithm of relative exposure) interval between them. Conversely, the log exposure interval between any two exposures is determined by finding the logarithm of their ratio.

As can be seen in Figure 47, the slope (or steepness) of the characteristic curves is continuously changing throughout the length of the curves. The effects of this change of slope on detail visibility are more completely explained on pages 136 to 139. It will suffice at this point to give a qualitative outline of these effects. For example, two slightly different thicknesses in the object radiographed transmit slightly different exposures to the film. These two exposures have a certain small log E interval between them, that is, have a certain ratio. The difference in the densities corresponding to the two exposures depends on just where on the characteristic curve they fall, and the steeper the slope of the curve, the greater is this density difference. For example, the curve of Film Z (Figure 47), is steepest in its middle portion. This means that a certain log E interval in the middle of the curve corresponds to a greater density difference than the same log E interval at either end of the curve. In other words, the film contrast is greatest where the slope of the characteristic curve is greatest. For Film Z, as has been pointed out, the region of greatest slope is in the central part of the curve. For Films X and Y, however, the slope—and hence the film contrast—continuously increases throughout the useful density range. The curves of most industrial x-ray films are similar to those of Films X and Y.

USE OF THE CHARACTERISTIC CURVE

The characteristic curve can be used to solve quantitative problems arising in radiography, in the preparation of technique charts, and in radiographic research. Ideally, characteristic curves made under the radiographic conditions actually encountered should be used in solving practical problems. However, it is not always possible to

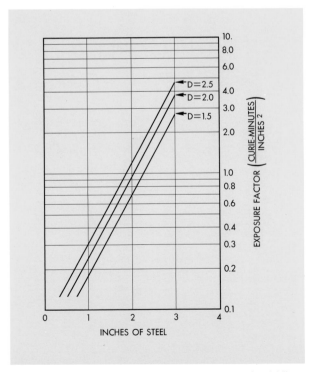

Figure 46—Typical gamma-ray exposure chart for iridium 192, based upon the use of Film X (Figure 47, for example).

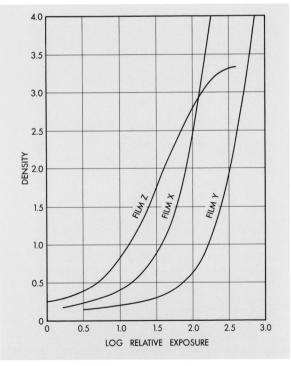

Figure 47—Characteristic curves of three typical x-ray films, exposed between lead foil screens.

produce characteristic curves in a radiographic department, and curves prepared elsewhere must be used. Such curves prove adequate for many purposes although it must be remembered that the shape of the characteristic curve and the speed of a film relative to that of another depend strongly on developing conditions. The accuracy attained when using "ready-made" characteristic curves is governed largely by the similarity between the developing conditions used in producing the characteristic curves and those for the films whose densities are to be evaluated.

A few examples of the quantitative use of characteristic curves are worked out on following pages. In the examples below, D is used for density and log E for logarithm of relative exposure.

Example 1: Suppose a radiograph made on Film Z (see Figure 48) with an exposure of 12 mA-min has a density of 0.8 in the region of maximum interest. It is desired to increase the density to 2.0 for the sake of the increased contrast there available.

① Log E at D = 2.0 is 1.62
② Log E at D = 0.8 is 1.00
③ Difference in log E is 0.62
  Antilogarithm of this difference is 4.2

Therefore, the original exposure is multiplied by 4.2 giving 50 mA-min to produce a density of 2.0.

Example 2: Film X has a higher contrast than Film Z at D = 2.0 (see Figure 49) and also a finer grain. Suppose that, for these reasons, it is desired to make the radiograph on Film X with a density of 2.0 in the same region of maximum interest.

④ Log E at D = 2.0 for Film X is 1.91
⑤ Log E at D = 2.0 for Film Z is 1.62
⑥ Difference in log E is 0.29
  Antilogarithm of this difference is 1.95

Therefore, the exposure for D = 2.0 on Film Z is multiplied by 1.95 giving 97.5 mA-min, for a density of 2.0 on Film X.

GRAPHICAL SOLUTIONS TO SENSITOMETRIC PROBLEMS

Many of the problems in the foregoing section can be solved graphically. One method involves the use of a transparent overlay (Figure 50) superimposed on a characteristic curve. Another method

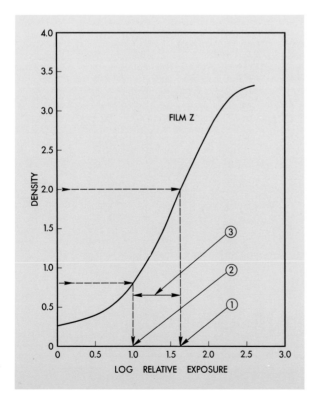

Figure 48—(For use with Example 1, above.) Circled numerals key corresponding items here and in Example 1.

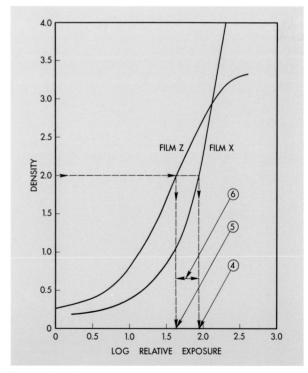

Figure 49—(For use with Example 2, above.) Characteristic curves of two x-ray films exposed with lead foil screens. Circled numerals key corresponding items in the figure and in Example 2.

involves the use of a nomogram-like chart (Figure 54, page 57). In general, the overlay method requires less arithmetic but more equipment than the nomogram method. The nomogram method usually requires somewhat more arithmetic but no equipment other than a diagram similar to Figure 54 and a ruler or straightedge.

Graphical solutions of either type are often sufficiently accurate for the purposes of practical industrial radiography.

Overlay Methods

An example of a transparent overlay is shown in Figure 50. The numbers on the horizontal line are exposure values. They can be taken, for example, to be milliampere-minutes, milliampere-seconds, curie-minutes, curie-hours or an exposure factor (see page 26). Further, all numbers on the line can be multiplied by the same value, without affecting the use of the device. For instance, by multiplying by 10, the scale can be made to go from 10 to 10,000 (rather than from 1 to 1,000) of whatever exposure unit is convenient. Note that the overlay must be made to fit the characteristic curves with which it is

to be used, since it is essential for the horizontal scales of both characteristic curves and overlay to agree.

The use of the overlay will be demonstrated by solving again some of the same problems used as illustrations in the foregoing section. Note that the vertical lines on the overlay must be parallel to the vertical lines on the graph paper of the characteristic curve, and the horizontal line must be parallel to the horizontal lines on the graph paper.

Example 1: Suppose a radiograph made on Film Z with an exposure of 12 mA-min has a density of 0.8 in the region of maximum interest. It is desired to increase the density to 2.0 for the sake of the increased contrast there available.

Locate the intersection of the line for the original density of 0.8 with the characteristic curve of Film Z (point A, Figure 51). Superimpose the transparent overlay on the curve, so that the vertical line for the original exposure—12 mA-min—passes through point A and the horizontal line overlies the line for the desired final density of 2.0. The new exposure, 50 mA-min, is read at the intersection of the charac-

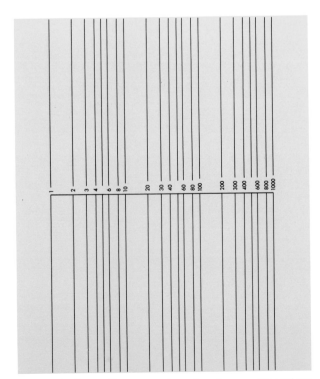

Figure 50—Form of transparent overlay of proper dimensions for use with characteristic curves of Figure 47 in solution of exposure problems. The use of the overlay and curves is demonstrated on pages 55 to 57.

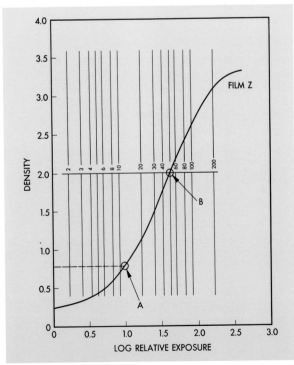

Figure 51—Characteristic curve of Film Z from Figure 47. Transparent overlay of Figure 50 positioned for the graphical solution of Example 1, above.

teristic curve with the horizontal line of the overlay (point B, Figure 51).

The method of solution would be the same if the new density were lower rather than higher than the old. The vertical line corresponding to the old exposure would pass through the characteristic curve at the point of the old density. The horizontal line of the overlay would pass through the desired new density. The new exposure would be read at the intersection of the characteristic curve and the horizontal line of the overlay.

Example 2: Film X has a higher contrast at D = 2.0 than Film Z and also has a finer grain. Suppose that, for these reasons, it is desired to make the aforementioned radiograph on Film X with a density of 2.0 in the same region of maximum interest.

Superimpose the overlay on the characteristic curve so that the horizontal line coincides with the horizontal line for a density of 2.0, and position the overlay from left to right so that the curve for Film Z cuts the line at the original exposure of 50 mA-min (point C, Figure 52). Read the new exposure of 97.5 mA-min at the point at which the curve

for Film X cuts the horizontal line (point D, Figure 52).

Example 3: The types of problems given in Examples 1 and 2 are often combined in actual practice. Suppose, for example, that a radiograph was made on Film X with an exposure of 20 mA-min and that a density of 1.0 was obtained. Then suppose that a radiograph at the same kilovoltage but on Film Y at a density of 2.5 is desired for the sake of the higher contrast and the lower graininess obtainable. The problem can be solved in a single step.

Locate the intersection of the original density of 1.0 and the characteristic curve of Film X (point E, Figure 53). Position the overlay so that the vertical line for the original exposure—20 mA-min— passes through point E and the horizontal line coincides with the line for the new density of 2.5. The new exposure of 220 mA-min is read from point F (Figure 53), the intersection of the horizontal line and the characteristic curve of Film Y.

Nomogram Methods

In Figure 54, the scales at the far left and far right are

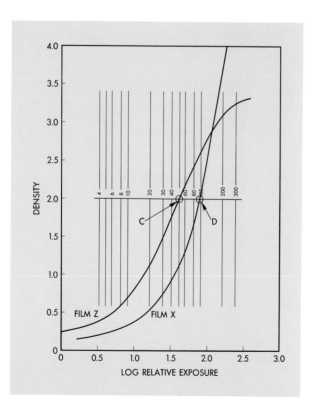

Figure 52—Characteristic curves of Films X and Z from Figure 47. Transparent overlay positioned for the graphical solution of Example 2, above.

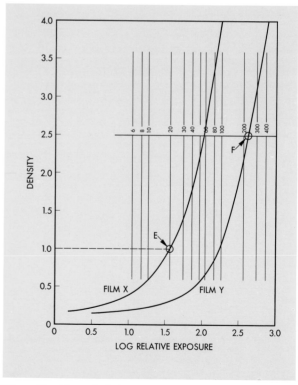

Figure 53—Characteristic curves of Films X and Y from Figure 47. Transparent overlay positioned for the graphical solution of Example 3, above.

relative exposure values. They do not represent milliampere-minutes, curie-hours or any other exposure unit; they are to be considered merely as multiplying (or dividing) factors, the use of which is explained below. Note, also, that these scales are identical, so that a ruler placed across them at the same value will intersect the vertical lines, in the center of the diagram, at right angles.

On the central group of lines, each labeled with the designation of a film whose curve is shown in Figure 47, the numbers represent *densities*.

The use of Figure 54 will be demonstrated by a re-solution of the same problems used as illustrations in both of the preceding sections. Note that in the use of the nomogram, the straightedge must be placed so that it is at right angles to all the lines—that is, so that it cuts the outermost scales on the left and the right at the same value.

Example 1: Suppose a radiograph made on Film Z (see Figure 47) with an exposure of 12 mA-min has a density of 0.8 in the region of maximum interest. It is desired to increase the density to 2.0 for the sake of the increased contrast there available.

Place the straightedge across Figure 54 so that it cuts the Film Z scale at 0.8. The reading on the outside scales is then 9.8. Now move the straightedge upward so that it cuts the Film Z scale at 2.0; the reading on the outside scales is now 41. The original exposure (12 mA-min) must be multiplied by the ratio of these two numbers—that is, by 41/9.8 = 4.2. Therefore, the new exposure is 12 × 4.2 mA-min or 50 mA-min.

Example 2: Film X has a higher contrast than Film Z at D = 2.0 (see Figure 47) and also lower graininess. Suppose that, for these reasons, it is desired to make the aforementioned radiograph on Film X with a density of 2.0 in the same region of maximum interest.

Place the straightedge on Figure 54 so that it cuts the scale for Film Z at 2.0. The reading on the outside scales is then 41, as in Example 1. When the straightedge is placed across the Film X scale at 2.0, the reading on the outside scale is 81. In the previous example the exposure for a density of 2.0 on Film Z was found to be 50 mA-min. In order to give a density of 2.0 on Film X, this exposure must be multiplied by the ratio of the two scale readings just found—81/41 = 1.97. The new exposure is therefore 50 × 1.97 or 98 mA-min.

Example 3: The types of problems given in Examples 1 and 2 are often combined in actual practice. Suppose, for example, that a radiograph was made on Film X (Figure 47) with an exposure of

20 mA-min and that a density of 1.0 was obtained. A radiograph at the same kilovoltage on Film Y at a density of 2.5 is desired for the sake of the higher contrast and the lower graininess obtainable. The problem can be solved graphically in a single step.

The reading on the outside scale for D = 1.0 on Film X is 38. The corresponding reading for D = 2.5 on Film Y is 420. The ratio of these is 420/38 = 11, the factor by which the original exposure must be multiplied. The new exposure to produce D = 2.5 on Film Y is then 20 × 11 or 220 mA-min.

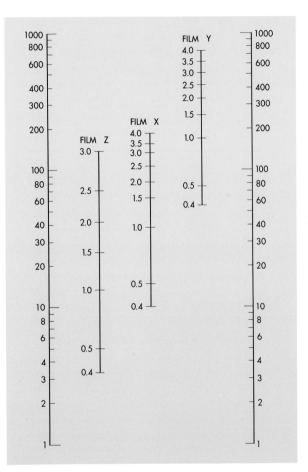

Figure 54—Typical nomogram for solution of exposure calculations. The uses of the diagram are explained on pages 56 and 57.

SLIDING SCALES FOR EXPOSURE CHARTS

An exposure chart is an exceedingly useful radiographic tool. However, as pointed out on page 51, it has the limitations of applying only to a specific set of radiographic conditions. These are:

1. The x-ray machine used

2. A certain source-film distance

3. A particular type of film

4. Processing conditions used

5. The film density on which the chart is based

6. The type of screens (if any) that are used

Only if the conditions used in practice agree in every particular with those used in the production of the exposure chart can exposures be read directly from the chart. If one or more of the conditions are changed, a correction factor must be applied to the exposure as determined from the chart. Correction factors to allow for differences between one x-ray machine and another, or between one type of screen and another, are best determined by experiment—often a new exposure chart must be made. Changes in the other four conditions, however, can in many cases be calculated, making use of the characteristic curve of the films involved (see pages 53 and 54) or of the inverse square law (page 23).

Numerical work involved in these corrections can often be avoided by the use of sliding scales affixed to the exposure chart. The preparation of these sliding scales will be facilitated (1) if the exposure chart has a logarithm of exposure (or of relative exposure) scale along one vertical boundary, as shown in Figures 44 and 46, and (2) if the horizontal (exposure) lines are available on a transparent overlay (Figure 55), on which the spacing of the lines corresponds to those in Figure 44. These overlays can be made by tracing from the exposure chart involved onto exposed, fixed-out, x-ray film or any stiff transparent plastic material.

Note that arrowheads are printed on both Figure 55 and Figure 56. When these coincide, the exposures read from the chart are those corresponding to the conditions under which the chart was made.

The material that follows describes the technique of modifying the exposure chart of Figure 44, to allow for changes in radiographic conditions 2 through 5 above. The sections below are numbered to correspond to the list on page 51.

2. *Source-film distance.* Since the intensity of the radiation varies inversely with the square of the source-film distance (page 23), the exposure must vary directly with the square of the distance (page 46) if a constant density on the radiograph is to be maintained. If source-film distance is to be changed, therefore, the exposure scale on the chart must be shifted vertically a distance that is in accord with the law.

The chart of Figure 44 was made using a 40-inch source-film distance. Assume that it is desired to make the chart applicable to distances of

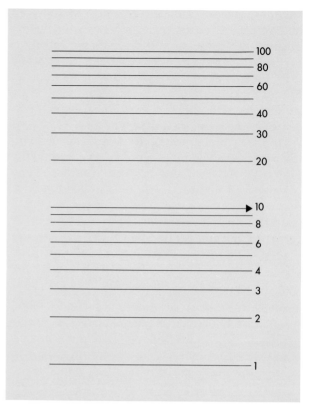

Figure 55—Pattern of transparent overlay for exposure chart of Figure 44.

30 and 60 inches as well. Note that column 2 of the tabulation below is calculated from the inverse square law *and that the ratio is taken so that it is always greater than 1.* (This is done merely for convenience, it being easier to work with logarithms of numbers greater than 1 than with logarithms of decimal fractions.) The logarithms in column 3 can be found by reference to Figure 43, the use of a slide rule, or the use of a table of logarithms.

Distance	Intensity Ratio (relative to 40″)	Logarithm of Intensity Ratio
30″	1.78*	0.25
60″	2.25†	0.35

*Intensity greater than that at 40 inches by this factor.
†Intensity less than that at 40 inches by this factor.

A mark is put on the margin of the exposure chart a log E *interval* of 0.25 above the printed arrow. When the transparent overlay is displaced upward to this position, exposures for a 30-inch focus-film distance can be read directly. Similarly, a mark is put a log E *interval* of 0.35 below the printed arrow. The overlay, in this position, gives the exposures required at a source-film distance of 60 inches. In Figure 57,

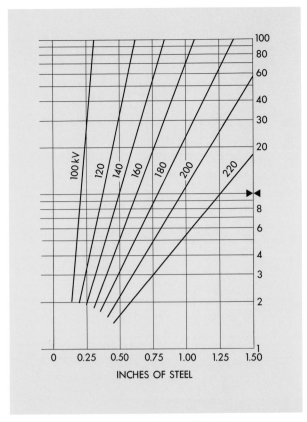

Figure 56—Transparent overlay positioned over exposure chart in such a way as to duplicate Figure 44. Thus, it applies to Film X (Figure 47) a density of 1.5 and a source-film distance of 40 inches.

the exposure chart and overlay are shown in this position.

3. *Film type.* Changes required by the use of a film different from that for which the exposure chart was prepared can be made by a somewhat similar procedure. Using the characteristic curve shown in Figure 47 and the method described in Example 2, pages 54, 56 or 57, it can be found that Film Y requires four times more exposure than does Film X to produce a density of 1.5. The logarithm of 4.0 is 0.60 (see Figure 43). A mark is put on the margin of the exposure chart a log exposure *interval* of 0.60 below the printed arrow. When the transparent overlay is in this position, exposures for Film Y can be read directly. Figure 58 shows this arrangement.

If the new film were faster than the one for which the chart was prepared, the same general procedure would be followed. The relative exposure required for the new film would be taken *so that it was greater than 1*, and the logarithm of this number would indicate the log E *interval* by

which the new mark would be placed *above* the printed arrow on the chart.

4. *Changes in processing conditions.* It is difficult to make simple corrections for gross changes in processing conditions, because such changes affect both the speed and the contrast of the film, often to a marked degree (see pages 141 to 143). Large departures from recommended times and temperatures should be avoided for the sake of both economy and radiographic quality.

However, in manual processing of some industrial x-ray films, the development time may be increased from, for example, 5 to 8 minutes at 68°F. The longer development may result in speed increases that are frequently useful in practice, with little or no change in the shape of the characteristic curve—that is, with little or no change in film contrast. Exposure charts of the types shown in Figures 44 and 54 can be made to apply to both development times. Assume, for instance, that increasing the development time of Film X from 5 to 8 minutes at 68°F results in a 20 percent gain in speed. The ratio of exposures required to achieve the same density would be 1:1.2. The logarithm of 1.2 is 0.08. A mark at log E *interval* of 0.08 above the arrow printed on the exposure chart would allow the overlay to be positioned so that exposures for a development time of 8 minutes could be read directly.

5. *Film density.* Exposure charts apply only to a single density of the processed radiograph. However, by the use of data from the characteristic curve of the film, it is possible to supply a sliding scale that can make the exposure chart applicable to any densities desired.

Figure 44 was drawn for a density of 1.5. Let us assume that it is desired to make this chart applicable to densities of 1.0, 2.0, and 3.0 also. The second column of the tabulation below is obtained from the characteristic curve of Film X in Figure 47.

Density	Log Relative Exposure	Difference of Log Relative Exposure
1.0	1.57	+ 0.22
1.5	1.79	0
2.0	1.92	− 0.13
3.0	2.10	− 0.31

The values in column 3 are the differences between the logarithm of relative exposure required to produce the density for which the chart was originally drawn (in this case 1.5) and the logarithm of exposure required for the desired density. They represent the log E *intervals* that the "exposure" grid of the exposure chart must be shifted up or down to give exposures which will result in the lower or higher densities. Plus signs indicate that the added marks on the margin should be above that printed on the chart; minus signs, that the added marks are to be below. Figure 59 shows the chart with the transparent overlay positioned to read directly exposures required to give a density of 2.0.

ESTIMATING EXPOSURES FOR MULTITHICKNESS SPECIMENS

A minimum acceptable density for radiographs is often specified, not because of any virtue in the particular density, but because the slope of the characteristic curve (and hence the film contrast) below a certain point is too low for adequate rendition of detail. Similarly, a maximum acceptable density is often designated because, either as with Film Z (Figure 47), the film contrast is lower at high densities or detail cannot be seen on the available illuminators if density is above a certain value.

The problem of radiographing a part having several thicknesses is one of using the available density range most efficiently. In other words, the kilovoltage and exposure should be adjusted so that the image of the thinnest part has the maximum acceptable density, and the thickest has the minimum. Exposure charts alone, although adequate for the radiography of uniform plates, can serve only as rough guides for articles having considerable variation in thickness. Previous experience is a guide, but even when a usable radiograph has been obtained, the question remains as to whether or not it is the best that could be achieved.

A quantitative method for finding such exposures combines information derived from the exposure chart and the characteristic curve of the film used. The procedure is outlined below:

Assume that 1.0 is the lowest acceptable density on Film X (Figure 47) and that 3.5 is the highest. As shown in Figure 60, this density interval corresponds to a certain log exposure interval, in this case 0.63.

The antilog of 0.63 is 4.3, which means that 4.3 times more exposure is required to produce a density of 3.5 than of 1.0. It is therefore desired that the thinnest portion of the object to be radiographed transmit exactly 4.3 times more radiation than the thickest part, so that with the proper adjustment of radiographic exposure all parts of the object will be rendered within the density range 1.0 to 3.5. The ratios of x-ray intensities transmitted by different portions of the object will depend on kilovoltage;

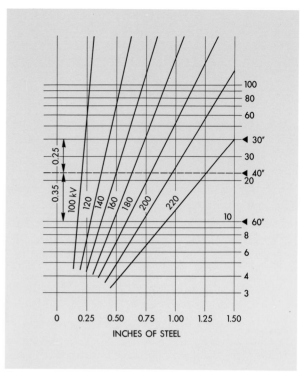

Figure 57—Overlay positioned so as to make exposure chart of Figures 44 and 54 apply to a source-film distance of 60 inches, rather than 40 inches. See No. 2, Source-film distance, page 58.

examination of the exposure chart of the x-ray machine reveals the proper choice of kilovoltage. For example, in the chart shown in Figure 61, the 180 kV line shows that a thickness range of about ⅞ to about 1¼ inches of steel corresponds to an exposure ratio of 35 mA-min to 8 mA-min, or 4.3, which is the ratio required. The next problem is to determine the radiographic exposure needed. The chart shown in Figure 61 gives the exposure to produce a density of 1.0 on Film X. Since it is desired to produce a density 1.0 under the thick section (1¼ inches), the exposure time would be 35 mA-min.

A simple means for applying this method to routine work is as follows:

Parallel lines are drawn on a transparent plastic sheet, such as a fixed-out x-ray film, in the manner shown in Figure 62. The spacing between the base line and the line immediately above is the log relative exposure interval for Film X between D = 1.0 and D = 3.5. *It is laid off to the same scale as the ordinate (vertical) scale of the exposure chart.* Similarly, the distance from the baseline to any other line parallel to it can be made to correspond to the log relative exposure interval for other density

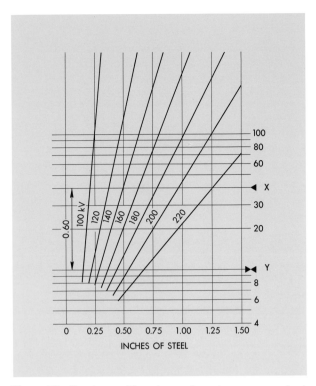

Figure 58—Overlay positioned so as to make exposure chart of Figures 44 and 54 apply to Film Y (Figure 47), rather than to Film X. See No. 3, Film type, page 59.

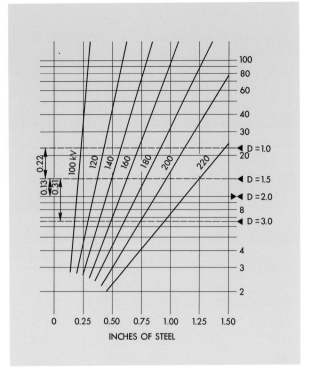

Figure 59—Overlay positioned so as to make exposure chart of Figures 44 and 55 apply to a density of 2.0, rather than 1.5. See No. 5, Film density, page 59.

limits and films. This transparent guide is moved up and down on the exposure chart with its lines parallel to the thickness axis. The two guidelines being used form a rectangle with the two vertical lines of the exposure chart which mark the thickness limits of the specimen. The correct kilovoltage is the one whose graph intersects diagonally opposite corners of the rectangle. If the film type used is the one for which the chart was prepared, the correct exposure is indicated at the intersection of the upper guide line with the exposure scale (point C, Figure 61). If a different film is used, a suitable correction factor obtained either from tables of relative speeds or by the method described in Example 2, pages 54, 56, or 57, must be applied to the exposure as determined from the chart.

If there is only one graph on a gamma-ray exposure chart, this procedure will indicate limiting thicknesses of material that can be radiographed within the prescribed density limits.

On a chart of the type shown in Figure 46, which has lines for various densities, the thickness range that can be radiographed in a single exposure can be read directly. For example, the same exposure

(exposure factor = 0.7) will give a density of 1.5 through 2 inches of steel and a density of 2.5 through about 1½ inches of steel.

USE OF MULTIPLE FILMS

If the chart shows that the thickness range is too great for a single exposure under any condition, it may be used to select two different exposures to cover the range. Another technique is to load the cassette with two films of different speed and expose them simultaneously, in which case the chart may be used to select the exposure. The log relative exposure range for two films of different speed, when used together in this manner, is the difference in log exposure between the value at the low-density end of the faster film curve and the high-density end of the slower film curve. Figure 47 shows that when Films X and Y are used the difference is 1.22, which is the difference between 1.57 and 2.79. It is necessary that the films be close enough together in speed so that their curves will have some "overlap" on the log E axis.

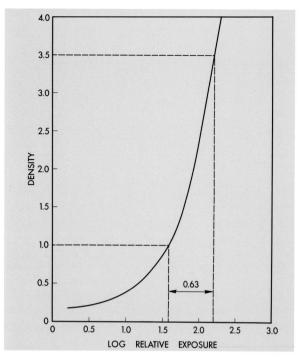

Figure 60—Characteristic curve of Film X (from Figure 47). Dotted lines show how the log E interval corresponding to a certain density interval (in this case 1.0 to 3.5) can be found.

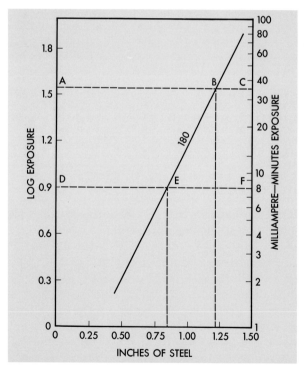

Figure 61—Abridged form of exposure chart derived from Figure 44, but showing exposures at 180 kV to produce a density of 1.0 on Film X (Figure 47). Dotted lines indicate the metal thickness corresponding to the log E interval of Figure 60. If the separation of lines ABC and DEF is maintained, they can be moved up and down the chart. They will then mark off a large number of thickness ranges on the various kilovoltage lines, all of which will completely fill the density range which has been assumed to be useful.

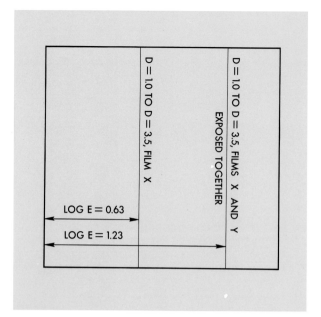

Figure 62—System of lines drawn on a transparent sheet to be used in connection with an exposure chart for estimating radiographic exposures for multithickness specimens.

LIMITATIONS OF EXPOSURE CHARTS

Although exposure charts are useful industrial radiographic tools, they must be used with some caution. They will, in most cases, be adequate for routine practice but they will not always show the precise exposure required to radiograph a given thickness to a particular density.

Several factors have a direct influence on the accuracy with which exposures can be predicted. Exposure charts are ordinarily prepared by radiographing a stepped wedge. Since the proportion of scattered radiation depends on the thickness of material and, therefore, on the distribution of the material in a given specimen, there is no assurance that the scattered radiation under different parts will correspond to the amount under the same thickness of the wedge. In fact, it is unreasonable to expect exact correspondence between scattering conditions under two objects the thicknesses of which are the same but in which the distribution of material is quite different. The more closely the distribution of metal in the wedge resembles that in the specimen the more accurately the exposure chart will serve its purpose. For example, a narrow wedge would approximate the scattering conditions for specimens containing narrow bars.

Although the lines of an exposure chart are normally straight, they should in most cases be curved

—concave downward. The straight lines are convenient approximations, suitable for most practical work, but it should be recognized that in most cases they are only approximations. The degree to which the conventionally drawn straight line approximates the true curve will vary, depending on the radiographic conditions, the quality of the exposing radiation, the material radiographed, and the amount of scattered radiation reaching the film.

In addition, time, temperature, degree of activity, and agitation of the developer are all variables which affect the shape of the characteristic curve and should therefore be standardized. When, in hand processing, the temperature or the activity of the developer does not correspond to the original conditions, proper compensation can be made by changing the time according to methods described on page 80. Automated processors should be carefully maintained and cleaned to achieve the most consistent results. In any event, the greatest of care should always be taken to follow the recommended processing procedures.

Factors Affecting Image Quality

RADIOGRAPHIC IMAGE QUALITY

RADIOGRAPHIC CONTRAST

DEFINITION

Subject Contrast

Affected by:

A Absorption differences in specimen (thickness, composition, density)

B Radiation wavelength (see pp 27, 40, 65, 66)

C Scattered radiation (see pp 38-43)

Reduced by:

1 Masks and diaphragms (see pp 39-40)

2 Filters (see pp 40-43, 44, 45)

3 Lead screens (see pp 30-32, 39)

4 Potter-Bucky diaphragm (see p 43)

Film Contrast

Affected by:

A Type of film (see pp 72, 136-139)

B Degree of development (type of developer; time and temperature of development; activity of developer; degree of agitation) (see pp 80-83, 141-143)

C Density (see pp 52-57, 136-139)

D Type of screens (fluorescent vs lead or none)

Geometric Factors

Affected by:

A Focal-spot size (see pp 15-20)

B Source-film distance (see pp 15-20)

C Specimen-film distance (see pp 15-20)

D Abruptness of thickness changes in specimen (see p 70)

E Screen-film contact (see pp 31, 36)

F Motion of specimen (see p 108)

Film Graininess, Screen Mottle Factors

Affected by:

A Type of film (see pp 66, 72)

B Type of screen (see pp 30, 35, 67)

C Radiation wavelength (see pp 67, 145)

D Development (see p 67)

CHAPTER 8

Radiographic Image Quality and Detail Visibility

Because the purpose of most radiographic inspections is to examine a specimen for inhomogeneity, a knowledge of the factors affecting the visibility of detail in the finished radiograph is essential. The preceding chart summary shows the relations of the various factors influencing image quality and radiographic sensitivity, together with page references to the discussion of individual topics. For convenience, a few important definitions will be repeated.

Radiographic sensitivity is a general or qualitative term referring to the size of the smallest detail that can be seen in a radiograph, or to the ease with which the images of small details can be detected. Phrased differently, it is a reference to the amount of information in the radiograph. Note that radiographic sensitivity depends on the combined effects of two independent sets of factors. One is radiographic contrast (the density difference between a small detail and its surroundings) and the other is definition (the abruptness and the "smoothness" of the density transition). See Figure 63.

Radiographic contrast between two areas of a radiograph is the difference between the densities of those areas. It depends on both subject contrast and film contrast. *Subject contrast* is the ratio of x-ray or gamma-ray intensities transmitted by two

selected portions of a specimen. (See Figure 64.) Subject contrast depends on the nature of the specimen, the energy (spectral composition, hardness, or wavelengths) of the radiation used, and the intensity and distribution of the scattered radiation, but is independent of time, milliamperage or source strength, and distance, and of the characteristics or treatment of the film.

Film contrast refers to the slope (steepness) of the characteristic curve of the film. It depends on the type of film, the processing it receives, and the density. It also depends on whether the film is exposed with lead screens (or direct) or with fluorescent screens. Film contrast is independent, for most practical purposes, of the wavelengths and distribution of the radiation reaching the film, and hence is independent of subject contrast.

Definition refers to the sharpness of outline in the image. It depends on the types of screens and film used, the radiation energy (wavelengths, etc), and the geometry of the radiographic setup.

SUBJECT CONTRAST

On page 27 it was stated that subject contrast decreases as kilovoltage is increased. The decreasing slope (steepness) of the lines of the exposure chart

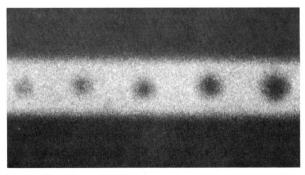 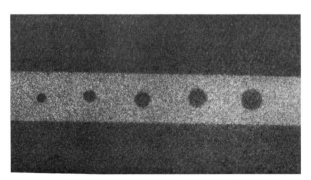

Figure 63—Advantage of higher radiographic contrast (left) is largely offset by poor definition. Despite lower contrast (right), better rendition of detail is obtained by improved definition.

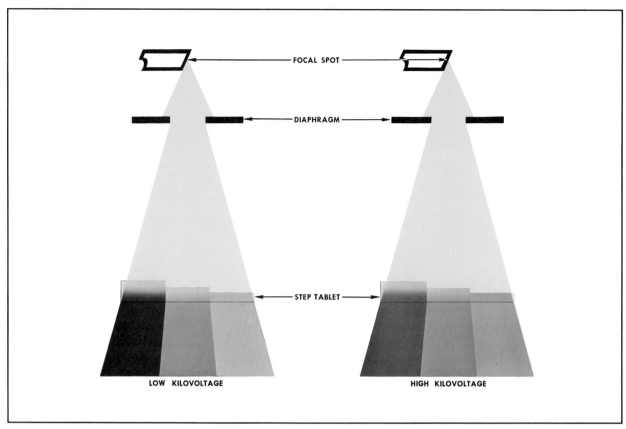

Figure 64—With the same specimen, the lower-kilovoltage beam (left) produces higher subject contrast than does the higher-kilovoltage beam (right).

(Figure 44, page 50) as kilovoltage increases illustrates the reduction of subject contrast as the radiation becomes more penetrating. For example, consider a steel part containing two thicknesses, ¾ inch and 1 inch, which is radiographed first at 160 kV and then at 200 kV.

kV	Inch Thickness	Exposure to give D = 1.5 mA-min	Relative Intensity	Ratio of Intensities
160	¾	18.5	3.8 ⎫	3.8
	1	70.0	1.0 ⎭	
200	¾	4.9	14.3 ⎫	2.5
	1	11.0	5.8 ⎭	

In the table above, column 3 shows the exposure in milliampere-minutes required to reach a density of 1.5 through each thickness at each kilovoltage. These data are from the exposure chart, Figure 44. It is apparent that the milliampere-minutes required to produce a given density at any kilovoltage are inversely proportional to the corresponding x-ray intensities passing through the different sections of the specimen. Column 4 gives these relative intensities for each kilovoltage. Column 5 gives the ratio of these intensities for each kilovoltage.

Column 5 shows that, at 160 kV, the intensity of the x-rays passing through the ¾-inch section is 3.8 times greater than that passing through the 1-inch section. At 200 kV, the radiation through the thinner portion is only 2.5 times that through the thicker. Thus, as the kilovoltage increases, the *ratio* of x-ray transmission of the two thicknesses decreases, indicating a lower subject contrast.

FILM CONTRAST

The dependence of film contrast on density must be kept in mind when considering problems of radiographic sensitivity. In general the contrast of radiographic films, except those designed for use with fluorescent screens, increases continuously with density in the usable density range. Therefore, for films that exhibit this continuous increase in contrast, the best density, or upper limit of density range, to use is the highest that can conveniently be viewed with the illuminators available. Adjustable high-intensity illuminators that greatly increase the maximum density that can be viewed are commercially available.

The use of high densities has the further advantage of increasing the range of radiation intensities which can be usefully recorded on a single film. This in turn permits, in x-ray radiography, the use of lower kilovoltage, with resulting increase in subject contrast and radiographic sensitivity.

Maximum contrast of screen-type films is at a density of about 2.0 (see page 138). Therefore, other things being equal, the greatest radiographic sensitivity will be obtained when the exposure is adjusted to give this density.

FILM GRAININESS, SCREEN MOTTLE

(See also Chapter 17)

The image on an x-ray film is formed by countless minute silver grains, the individual particles being so small that they are visible only under a microscope. However, these small particles are grouped together in relatively large masses which are visible to the naked eye or with a magnification of only a few diameters. These masses result in the visual impression called *graininess.*

All films exhibit graininess to a greater or lesser degree. In general, the slower films have lower graininess than the faster. Thus, Film Y (Figure 47) would have a lower graininess than Film X.

The graininess of all films increases as the penetration of the radiation increases, although the rate of increase may be different for different films. The graininess of the images produced at high kilovoltages makes the slow, inherently fine-grain films especially useful in the million- and multimillion-volt range. When sufficient exposure can be given, they are also useful with gamma rays.

The use of lead screens has no significant effect on film graininess. However, graininess is affected by processing conditions, being directly related to the degree of development. For instance, if development time is increased for the purpose of increasing film speed, the graininess of the resulting image is likewise increased. Conversely, a developer or developing technique that results in an appreciable decrease in graininess will also cause an appreciable loss in film speed. However, adjustments made in development technique to compensate for changes in temperature or activity of a developer will have little effect on graininess. Such adjustments are made to achieve the same degree of development as would be obtained in the fresh developer at a standard processing temperature, and therefore the graininess of the film will be essentially unaffected.

Another source of the irregular density in uniformly exposed areas is the screen mottle encountered in radiography with fluorescent screens (see page 35). The screen mottle increases markedly as hardness of the radiation increases. This is one of the factors that limits the use of fluorescent screens at high voltage and with gamma rays.

PENETRAMETERS

A standard test piece is usually included in every radiograph as a check on the adequacy of the radiographic technique. The test piece is commonly referred to as a penetrameter in North America and an Image Quality Indicator (IQI) in Europe. The penetrameter (or IQI) is made of the same material, or a similar material, as the specimen being radiographed, and is of a simple geometric form. It contains some small structures (holes, wires, etc), the dimensions of which bear some numerical relation to the thickness of the part being tested. The image of the penetrameter on the radiograph is permanent evidence that the radiographic examination was conducted under proper conditions.

Codes or agreements between customer and vendor may specify the type of penetrameter, its dimensions, and how it is to be employed. Even if penetrameters are not specified, their use is advisable, because they provide an effective check on the overall quality of the radiographic inspection.

Hole Type Penetrameters

The common penetrameter consists of a small rectangular piece of metal, containing several (usually three) holes, the diameters of which are related to the thickness of the penetrameter (Figure 65).

The ASTM (American Society for Testing and Materials) penetrameter contains three holes of diameters T, 2T, and 4T, where T is the thickness of the penetrameter. Because of the practical difficulties in drilling minute holes in thin materials, the minimum diameters of these three holes are 0.010, 0.020, and 0.040 inches, respectively. These penetrameters may also have a slit similar to the ASME penetrameter described on page 68. Thick penetrameters of the hole type would be very large, because of the diameter of the 4T hole. Therefore, penetrameters more than 0.180 inch thick are in the form of discs, the diameters of which are 4 times the thickness (4T) and which contain two holes of diameters T and 2T. Each penetrameter is identified by a lead number showing the thickness in thousandths of an inch.

The ASTM penetrameter permits the specification of a number of levels of radiographic sensitivity, depending on the requirements of the job. For example, the specifications may call for a radiographic sensitivity level of 2-2T. The first symbol (2) indicates that the penetrameter shall be 2 percent of the thickness of the specimen; the second (2T) indicates that the hole having a diameter twice the penetrameter thickness shall be visible on the finished radiograph. The quality level 2-2T is probably the one most commonly specified for routine radiography. However, critical components may require more rigid standards, and a level of 1-2T or 1-1T may be required. On the other hand, the radiography of less critical specimens may be satisfactory if a quality level of 2-4T or 4-4T is achieved. The more critical the radiographic examination—that is, the *higher* the level of radiographic sensitivity required—the *lower* the numerical designation for the quality level.

Some sections of the ASME (American Society of Mechanical Engineers) Boiler and Pressure Vessel Code require a penetrameter similar in general to the ASTM penetrameter. It contains three holes, one of which is 2T in diameter, where T is the penetrameter thickness. Customarily, the other two holes are 3T and 4T in diameter, but other sizes may be used. Minimum hole size is 1/16 inch. Penetrameters 0.010 inch, and less, in thickness also contain a slit 0.010-inch wide and 1/4-inch long. Each is identified by a lead number designating the thickness in thousandths of an inch.

Equivalent Penetrameter Sensitivity

Ideally, the penetrameter should be made of the same material as the specimen. However, this is sometimes impossible because of practical or economic difficulties. In such cases, the penetrameter may be made of a radiographically similar material—that is, a material having the same radiographic absorption as the specimen, but one of which it is easier to make penetrameters. Tables of radiographically equivalent materials have been published wherein materials having similar radiographic absorptions are arranged in groups. In addition, a penetrameter made of a particular material may be used in the radiography of materials having *greater* radiographic absorption. In such a case, there is a certain penalty on the radiographic testers, because they are setting for themselves more rigid radiographic quality standards than are actually required. The penalty is often outweighed, however, by avoidance of the problems of obtaining penetrameters of an unusual material or one of which it is difficult to make penetrameters.

In some cases, the materials involved do not appear in published tabulations. Under these cir-

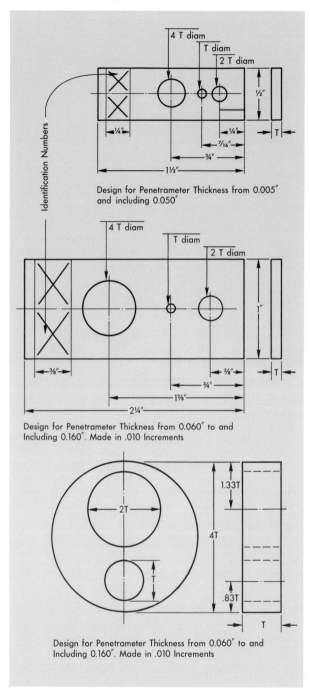

Design for Penetrameter Thickness from 0.005" and including 0.050"

Design for Penetrameter Thickness from 0.060" to and Including 0.160". Made in .010 Increments

Design for Penetrameter Thickness from 0.060" to and Including 0.160". Made in .010 Increments

Figure 65—American Society for Testing and Materials (ASTM) penetrameter (ASTM E 142-68).

cumstances the comparative radiographic absorption of two materials may be determined experimentally. A block of the material under test and a block of the material proposed for penetrameters, equal in thickness to the part being examined, can be radiographed side by side on the same film with the technique to be used in practice. If the density under the proposed penetrameter material is equal to or greater than the density under the specimen material, that proposed material is suitable for fabrication of penetrameters.

In practically all cases, the penetrameter is placed on the source side of the specimen—that is, in the least advantageous geometric position. In some instances, however, this location for the penetrameter is not feasible. An example would be the radiography of a circumferential weld in a long tubular structure, using a source positioned within the tube and film on the outer surface. In such a case a "film-side" penetrameter must be used. Some codes specify the film-side penetrameter that is equivalent to the source-side penetrameter normally required. When such a specification is not made, the required film-side penetrameter may be found experimentally. In the example above, a short section of tube of the same dimensions and materials as the item under test would be used to demonstrate the technique. The required penetrameter would be used on the source side, and a range of penetrameters on the film side. If the penetrameter on the source side indicated that the required radiographic sensitivity was being achieved, the image of the smallest visible penetrameter hole in the film-side penetrameters would be used to determine the penetrameter and the hole size to be used on the production radiographs.

Sometimes the shape of the part being examined precludes placing the penetrameter on the part. When this occurs, the penetrameter may be placed on a block of radiographically similar material of the same thickness as the specimen. The block and the penetrameter should be placed as close as possible to the specimen.

Wire Penetrameters

A number of other penetrameter designs are also in use. The German DIN (Deutsche Industrie-Norm) penetrameter (Figure 66) is one that is widely used. It consists of a number of wires, of various diameters, sealed in a plastic envelope that carries the necessary identification symbols. The image quality is indicated by the thinnest wire visible on the radiograph. The system is such that only three penetrameters, each containing seven wires, can cover a very wide range of specimen thicknesses. Sets of DIN penetrameters are available in aluminum, copper, and steel. Thus a total of nine penetrameters is sufficient for the radiography of a wide range of materials and thicknesses.

Comparison of Penetrameter Design

The hole type of penetrameter (ASTM, ASME) is, in a sense, a "go no-go" gauge; that is, it indicates whether or not a specified quality level has been attained but, in most cases, does not indicate whether the requirements have been exceeded, or by how much. The DIN penetrameter on the other hand is a series of seven penetrameters in a single unit. As such, it has the advantage that the radiographic quality level achieved can often be read directly from the processed radiograph.

On the other hand, the hole penetrameter can be made of any desired material but the wire penetrameter is made from only a few materials. Therefore, using the hole penetrameter, a quality level of 2-2T may be specified for the radiography of, for example, commercially pure aluminum and 2024 aluminum alloy, even though these have appreciably different compositions and radiation absorptions. The penetrameter would, in each case, be made of the appropriate material. The wire penetrameters, however, are available in aluminum but not in 2024 alloy. To achieve the same quality of radiographic inspection of equal thicknesses of these two materials, it would be necessary to specify different wire diameters—that for 2024 alloy would probably have to be determined by experiment.

Special Penetrameters

Special penetrameters have been designed for certain classes of radiographic inspection. An example is the radiography of small electronic components wherein some of the significant factors are the continuity of fine wires or the presence of tiny balls of solder. Special image quality indicators have been designed consisting of fine wires and small metallic spheres within a plastic block, the whole covered

Figure 66—DIN (German) penetrameter (German Standard DIN 54109).

on top and bottom with steel approximately as thick as the case of the electronic component.

Penetrameters and Visibility of Discontinuities

It should be remembered that even if a certain hole in a penetrameter is visible on the radiograph, a cavity of the same diameter and thickness may not be visible. The penetrameter holes, having sharp boundaries, result in an abrupt, though small, change in metal thickness whereas a natural cavity having more or less rounded sides causes a gradual change. Therefore, the image of the penetrameter hole is sharper and more easily seen in the radiograph than is the image of the cavity. Similarly, a fine crack may be of considerable extent, but if the x-rays or gamma rays pass from source to film along the thickness of the crack, its image on the film may not be visible because of the very gradual transition in photographic density. Thus, a penetrameter is used to indicate the quality of the radiographic technique and not to measure the size of cavity which can be shown.

In the case of a wire image quality indicator of the DIN type, the visibility of a wire of a certain diameter does not assure that a discontinuity of the same cross section will be visible. The human eye perceives much more readily a long boundary than it does a short one, even if the density difference and the sharpness of the image are the same.

VIEWING AND INTERPRETING RADIOGRAPHS

The examination of the finished radiograph should be made under conditions that favor the best visibility of detail combined with a maximum of comfort and a minimum of fatigue for the observer. To be satisfactory for use in viewing radiographs, an illuminator must fulfill two basic requirements. First, it must provide light of an intensity that will illuminate the areas of interest in the radiograph to their best advantage, free from glare. Second, it must diffuse the light evenly over the entire viewing area. The color of the light is of no optical consequence, but most observers prefer bluish white. An illuminator incorporating several fluorescent tubes meets this requirement and is often used for viewing industrial radiographs of moderate density.

For routine viewing of high densities, one of the commercially available high-intensity illuminators should be used. These provide an adjustable light source, the maximum intensity of which allows viewing of densities of 4.0 or even higher.

Such a high-intensity illuminator is especially useful for the examination of radiographs having a wide range of densities corresponding to a wide range of thicknesses in the object. If the exposure was adequate for the greatest thickness in the specimen, the detail reproduced in other thicknesses can be visualized with illumination of sufficient intensity.

The contrast sensitivity of the human eye (that is, the ability to distinguish small brightness differences) is greatest when the surroundings are of about the same brightness as the area of interest. Thus, to see the finest detail in a radiograph, the illuminator must be masked to avoid glare from bright light at the edges of the radiograph, or transmitted by areas of low density. Subdued lighting, rather than total darkness, is preferable in the viewing room. The room illumination must be such that there are no troublesome reflections from the surface of the film under examination.

Modern x-ray films for general radiography consist of an emulsion—gelatin containing a radiation-sensitive silver compound—and a flexible, transparent, blue-tinted base. Usually, the emulsion is coated on both sides of the base in layers about 0.0005 inch thick. (See Figures 67 and 68.) Putting emulsion on both sides of the base doubles the amount of radiation-sensitive silver compound, and thus increases the speed. At the same time, the emulsion layers are thin enough so that developing, fixing, and drying can be accomplished in a reasonable time. However, some films for radiography in which the highest detail visibility is required have emulsion on only one side of the base.

When x-rays, gamma rays, or light strike the grains of the sensitive silver compound in the emulsion, a change takes place in the physical structure of the grains. This change is of such a nature that it cannot be detected by ordinary physical methods. However, when the exposed film is treated with a chemical solution (called a *developer*), a reaction takes place, causing the formation of

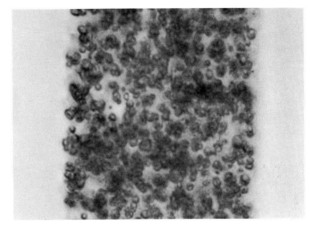

Figure 68—Cross section of the unprocessed emulsion on one side of an x-ray film. Note the large number of grains as compared to the developed grains of Figure 69.

black, metallic silver. It is this silver, suspended in the gelatin on both sides of the base, that constitutes the image. (See Figure 69.) The details of this process are discussed in greater length in Chapter 10.

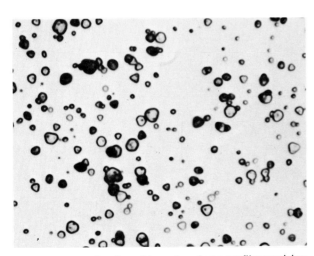

Figure 67—The silver bromide grains of an x-ray film emulsion (2,500 diameters). These grains have been dispersed to show their shape and relative sizes more clearly. In an actual coating, the crystals are much more closely packed.

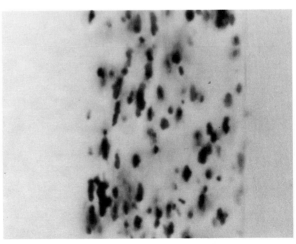

Figure 69—Cross section showing the distribution of the developed grains in an x-ray film emulsion exposed to give a moderate density.

Although an image may be formed by light and other forms of radiation, as well as by gamma rays or x-rays, the properties of the latter two are of a distinct character, and, for this reason, the sensitive emulsion must be different from those used in other types of photography.

SELECTION OF FILMS FOR INDUSTRIAL RADIOGRAPHY

As pointed out at the beginning of this book, industrial radiography now has many widely diverse applications. There are many considerations in obtaining the best radiographic results, for example: (1) the composition, shape, and size of the part being examined—and, in some cases, its weight and location as well; (2) the type of radiation used—whether x-rays from an x-ray machine or gamma rays from a radioactive material; (3) the kilovoltages available with the x-ray equipment; (4) the intensity of the gamma radiation; (5) the kind of information sought—whether it is simply an overall inspection or the critical examination of some especially important portion, characteristic, or feature; and (6) the resulting relative emphasis on definition, contrast, density, and the time required for proper exposure. All of these considerations are important in the determination of the most effective combination of radiographic technique and x-ray film.

The selection of a film for the radiography of any particular part depends on the thickness and material of the specimen and on the voltage range of the available x-ray machine. In addition, the choice is affected by the relative importance of high radiographic quality or short exposure time. Thus, an attempt must be made to balance these two opposing factors. As a consequence, it is not possible to present definite rules on the selection of a film. If high quality is the deciding factor, a slower and hence finer grained film should be substituted for a faster one—for instance, for the radiography of steel up to ¼ inch thick at 120-150 kV, Film Y (Figure 47) might be substituted for Film X. If short exposure times are essential, a faster film (or film-screen combination) can be used. For example, 1½-inch steel might be radiographed at 200 kV using fluorescent screens and a film particularly sensitive to blue light, rather than a direct exposure film with lead screens.

Figure 70 indicates the direction that these substitutions take. The "direct-exposure" films may be used with or without lead screens, depending upon the kilovoltage and upon the thickness and shape of the specimen. (See Chapter 5.)

Fluorescent intensifying screens must be used in radiography requiring the highest possible photographic speed (see page 33). The light emitted by the screens has a much greater photographic action than the x-rays either alone or combined with the emission from lead screens. To secure adequate exposure within a reasonable time, screen-type x-ray films sandwiched between fluorescent intensifying screens are often used in radiography of steel in thicknesses greater than about 2 inches at 250 kV, and more than about 3 inches at 400 kV.

FILM PACKAGING

Industrial x-ray films are available in a number of different types of packagings, each one ideally suited to particular classes of radiography.

Sheet Films

Formerly, x-ray films were available only in individual sheets, and this form is still the most popular

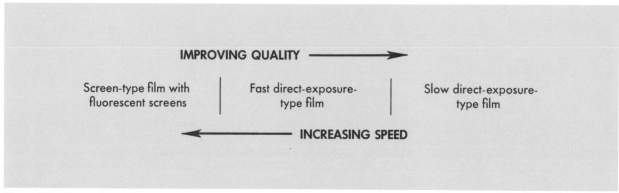

Figure 70—Change in choice of film, depending on relative emphasis on high speed or high radiographic quality.

packaging. Each sheet of film may be enclosed in an individual paper folder (interleaved). The choice between the interleaved and non-interleaved films is a matter of the user's preference. When no-screen techniques are used, the interleaving paper can be left on the film during exposure, providing additional protection to the film against accidental fogging by light or marking by moist fingertips. In addition, many users find the interleaving folders useful in filing the finished radiographs, protecting them against scratches and dirt during handling, and providing a convenient place for notes and comments about the radiograph.

Envelope Packing

Industrial x-ray films are also available in a form in which each sheet is enclosed in a folder of interleaving paper sealed in a lighttight envelope. The film can be exposed from either side without removing it from the envelope. A rip strip makes it easy to remove the film in the darkroom for processing. This form of packaging has the advantage that the time-consuming process of darkroom loading of cassettes and film holders is eliminated. The film is completely protected from finger marks and dirt until the time the film is removed from the envelope for processing.

Envelope Packing with Integral Lead Oxide Screens

The main feature of this type of packaging is that the sheet of film in an envelope is enclosed between two lead oxide screens which are in direct contact with the film. This form of packaging affords great convenience in material handling in an industrial x-ray department. As pointed out on page 33, it provides the advantage of cleanliness. This is particularly important where heavy inclusions in the specimen are significant. The use of film in this packaging prevents the images of such inclusions from being confused with artifacts caused by dust, cigarette ash, and the like being introduced between film and screen during darkroom handling. The time-consuming process of loading and unloading cassettes and film holders is avoided.

Roll Films

In the radiography of circumferential welds in cylindrical specimens, in the examination of the joints of a complete frame of an aircraft fuselage, and the like, long lengths of film permit great economies. The film is wrapped around the outside of the structure and the radiation source is positioned on the axis inside allowing the examination of the entire circumference to be made with a single exposure. Long rolls of film are also convenient for use in mechanized exposure holders for the repetitive radiography of identical specimens or for step-and-repeat devices in which radiation source and film holder move in synchronism along an extended specimen.

HANDLING OF FILM

X-ray film should always be handled carefully to avoid physical strains, such as pressure, creasing, buckling, friction, etc. The normal pressure applied in a cassette to provide good contacts is not enough to damage the film. However, whenever films are loaded in semiflexible holders and external clamping devices are used, care should be taken to be sure that this pressure is uniform. If a film holder bears against a few high spots, such as occur on an unground weld, the pressure may be great enough to produce desensitized areas in the radiograph. This precaution is particularly important when using envelope-packed films.

Marks resulting from contact with fingers that are moist or contaminated with processing chemicals, as well as crimp marks, are avoided if large films are always grasped by the edges and allowed to hang free. A convenient supply of clean towels is an incentive to dry the hands often and well. Use of envelope-packed films avoids these problems until the envelope is opened for processing. Thereafter, of course, the usual precautions must be observed.

Another important precaution is to avoid drawing film rapidly from cartons, exposure holders, or cassettes. Such care will help materially to eliminate objectionable circular or treelike black markings in the radiograph, the results of static electric discharges.

The interleaving paper should be removed before the film is loaded between either lead or fluorescent screens. When using exposure holders in direct exposure techniques, however, the paper should be left on the film for the added protection that it provides. At high voltage, direct-exposure techniques are open to the objection pointed out on page 33, that electrons emitted by the lead backing of the cassette or exposure holder may reach the film through the intervening paper or felt and record an image of this material on the film. This effect is avoided by the use of lead or fluorescent screens. In the radiography of light metals, direct-exposure techniques are the rule, and the paper folder should be left on interleaved film when loading it in the exposure holder.

The ends of a length of roll film factory-packed in a paper sleeve should be sealed in the darkroom with black pressure-sensitive tape. The tape should extend beyond the edges of the strip ¼ to ½ inch, to provide a positive lighttight seal.

IDENTIFYING RADIOGRAPHS

Because of their high absorption, lead numbers or letters affixed to the subject furnish a simple means of identifying radiographs. They may also be used as reference marks to determine the location of discontinuities within the specimen. Such markers can be conveniently fastened to the object with adhesive tape. A code can be devised to minimize the amount of lettering needed. Lead letters are commercially available in a variety of sizes and styles. The thickness of the letters chosen should be great enough so that their image is clearly visible on exposures with the most penetrating radiation routinely used. Under some circumstances it may be necessary to put the lead letters on a radiation-absorbing block so that their image will not be "burned out." The block should be considerably larger than the legend itself.

SHIPPING OF UNPROCESSED FILMS

If unprocessed film is to be shipped, the package should be carefully and conspicuously labeled, indicating the contents, so that the package may be segregated from any radioactive materials. It should further be noted that customs inspection of shipments crossing international boundaries sometimes includes fluoroscopic inspection. To avoid damage from this cause, packages, personal baggage, and the like containing unprocessed film should be plainly marked, if possible, and the attention of inspectors drawn to their contents.

STORAGE OF UNPROCESSED FILM

With x-rays generated up to 200 kV, it is feasible to use storage compartments lined with a sufficient thickness of lead to protect the film. At higher kilovoltages, protection becomes increasingly difficult; hence, film should be protected not only by the radiation barrier for protection of personnel but also by increased distance from the source.

At 100 kV, a ⅛-inch thickness of lead should normally be adequate to protect film stored in a room adjacent to the x-ray room if the film is not in the line of the direct beam. At 200 kV, the lead thickness should be increased to ¼ inch.

With million-volt x-rays, films should be stored beyond the concrete or other protective wall at a distance at least five times farther from the x-ray tube than the area occupied by personnel. The storage period should not exceed the times recommended by the manufacturer.

Medical x-ray films should be stored at approximately 12 times the distance of the personnel from the million-volt x-ray tube, for a total storage period *not exceeding two weeks.*

In this connection, it should be noted that the shielding requirements for films given in National Bureau of Standards Handbook 76 "Medical X-Ray Protection Up to Three Million Volts" and National Bureau of Standards Handbook 93 "Safety Standard for Non-Medical X-Ray and Sealed Gamma-Ray Sources, Part 1 General" are *not* adequate to protect the faster types of x-ray films in storage.

Gamma Rays

When radioactive material is not in use, the lead container in which it is stored helps provide protection for film. In many cases, however, the storage container for gamma-ray source will not provide completely satisfactory protection to stored x-ray film. In such cases, the emitter and stored film should be separated by a sufficient distance to prevent fogging. The conditions for the safe storage of x-ray film in the vicinity of gamma-ray emitters are given in Tables VII and VIII.

These show the necessary emitter-film distances and thicknesses of lead that should surround the various gamma-ray emitters to provide protection of stored film. These recommendations allow for a slight but harmless degree of fog on films when stored for the recommended periods.

The lead thicknesses and distances in Tables VII and VIII are considered the minimum tolerable.

To apply the cobalt 60 table to radium, values for source strength should be multiplied by 1.6 to give the *grams* of radium in a source which will have the same gamma-ray output and hence will require the same lead protection. This table can be extended to larger or smaller source sizes very easily. The half-value layer, in lead, for the gamma rays of radium or cobalt 60 is about ½ inch. Therefore, if the source strength is doubled or halved, the lead protection should be increased or decreased by ½ inch.

The table can also be adapted to storage times longer than those given in the tabulation. If, for example, film is to be stored in the vicinity of cobalt 60 for twice the recommended time, the protection recommendations for a source *twice* as large as the actual source should be followed.

TABLE VII—COBALT 60 STORAGE CONDITIONS FOR FILM PROTECTION

Source Strength in Curies	1	5	10	25	50	100
Distance from Film Storage in Feet	LEAD SURROUNDING SOURCE IN INCHES*					
25	5.5	7.0	7.5	8.0	8.5	9.0
50	4.5	6.0	6.5	7.0	7.5	8.0
100	3.5	5.0	5.5	6.0	6.5	7.0
200	2.5	4.0	4.5	5.0	5.5	6.0
400	1.5	3.0	3.5	4.0	4.5	5.0

*Lead thicknesses rounded off to nearest half-value layer.

TABLE VIII—IRIDIUM 192 STORAGE CONDITIONS FOR FILM PROTECTION

Output R/hr at 1 Meter	1	2	5	10	25	50
Source Strength in Curies	2	5	12.5	25	75	150
Distance from Film Storage in Feet	LEAD SURROUNDING SOURCE IN INCHES*					
25	1.70	1.85	2.15	2.35	2.50	2.70
50	1.35	1.50	1.85	2.00	2.15	2.35
100	1.00	1.15	1.50	1.70	1.85	2.00
200	0.70	0.85	1.15	1.35	1.50	1.70
400	0.35	0.50	0.85	1.00	1.15	1.35

*Lead thicknesses rounded off to nearest half-value layer.

Iridium 192 has a high absorption for its own gamma radiation. This means that the external radiation from a large source is lower, per curie of activity, than that from a small source. Therefore, protection requirements for an iridium 192 source should be based on the radiation output, in terms of roentgens per hour at a known distance. The values of source strength, in curies, are merely a rough guide, and should be used only if the radiation output of the source is unknown. The table above can be extended to sources having higher or lower radiation outputs than those listed. The half-value layer of iridium 192 radiation in lead is about 1/6 inch. Therefore, if the radiation output is doubled or halved, the lead thicknesses should be respectively increased or decreased by 1/6 inch.

Tables VII and VIII are based on the storage of a particular amount of radioactive material in a single protective lead container. The problem of protection of film from gamma radiation becomes more complicated when the film is exposed to radiation from several sources, each in its own housing. Assume that a radiographic source is stored under the conditions required by Table VII or VIII (for example, a 50-curie cobalt 60 source, in a 6.5-inch lead container 100 feet from the film storage). This combination of lead and distance would adequately protect the film from the gamma radiation for the storage times given in the tables. However, if a second source, identical with the first and in a similar container, is stored alongside the first, the radiation level at the film will be doubled. Obviously, then, if there are several sources in separate containers, the lead protection around each or the distance from the sources to the film must be increased over the values given in the tables.

The simplest method of determining the film protection required for several sources is as follows. Multiply the actual total strength of the source in each container by the number of separate containers. Then use these assumed source strengths to choose lead thicknesses and distances from Tables VII and VIII, and apply the values so found for the protection around each of the actual sources. For instance, assume that in a particular radiographic department there are two source containers, both at 100 feet from the film storage area. One container holds 50 curies of cobalt 60 and the other an iridium 192 source whose output is 5 roentgens per hour at 1 metre (5 rhm). Since there are two sources, the 50 curies of cobalt 60 will require the protection needed for a "solitary" 100-curie source, and the iridium 192 source will need the same protection as if a source whose output is 10 rhm were alone irradiating the stored film. The thicknesses of lead needed are shown to be 7.0 inches for the 50 curies of cobalt 60 (Table VII) and 1.7 inches for the iridium 192 whose emission is 5 rhm (Table VIII).

This method of determining the protective requirements when multiple sources must be considered is based on two facts. First, if several sources, say four, simultaneously irradiate stored film, the exposure contributed by each must be only one-quarter that which would be permissible if each source were acting alone—in other words, the gamma-ray attenuation must be increased by a factor of four. Second, any combination of source strength, lead thickness, and distance given in Tables VII and VIII results in the same gamma-ray dose rate—about 0.017 mr per hour—being delivered to the film location. Thus, to determine conditions which would reduce the radiation from a particular source to one-quarter the value on which Tables VII and VIII are based, it is only necessary to use the conditions that are set up for a source four times the actual source strength.

Heat, Humidity and Fumes

During packaging, most x-ray films are enclosed in a moistureproof container which is hermetically sealed and then boxed. As long as the seal is unbroken, the film is protected against moisture and fumes. Because of the deleterious effect of heat, all films should be stored in a cool, dry place and ordered in such quantities that the supply of film on hand is renewed frequently.

Under no circumstances should opened boxes of film be left in a chemical storage room or in any location where there is leakage of illuminating gas or any other types of gases, or where there is a possibility of contact with formalin vapors, hydrogen sulfide, ammonia or hydrogen peroxide.

Packages of sheet film should be stored on edge—that is, with the plane of the film vertical. They should *not* be stacked with the boxes horizontal because the film in the bottom boxes can be damaged by the impact of stacking or by the weight of boxes above. In addition, storage of the boxes on edge makes it simpler to rotate the inventory —that is, to use the older films first.

STORAGE OF EXPOSED AND PROCESSED FILM

Archival Keeping

Many factors affect the storage life of radiographs. One of the most important factors is residual thiosulfate (from the fixer chemicals) left in the radiograph after processing and drying. For archival storage,* ANSI PH1.41 specifies the amount of residual thiosulfate (as determined by the methylene blue test) to be a maximum level of 2 micrograms/cm² on each side of coarse-grain x-ray films. For short-term storage requirements, the residual thiosulfate content can be at a higher level, but this level is not specified. Washing of the film after development and fixing, therefore, is most important. The methylene blue test and silver densitometric test are laboratory procedures to be performed on *clear areas* of the processed film.

The following ANSI documents† may be used as an aid in determining storage conditions.

1. ANSI PH1.41, Specifications for Photographic

76

Film for Archival Records, Silver Gelatin Type on Polyester Base.

2. ANSI PH1.43, Practice for Storage of Processed Safety Photographic Film.

3. ANSI PH4.20, Requirements for Photographic Filing Enclosures for Storing Processed Photographic Films, Plates, and Papers.

4. ANSI PH4.8, Methylene Blue Method for Measuring Thiosulfate and Silver Densitometric Method for Measuring Residual Chemicals in Film, Plates, and Papers.

5. ANSI N45.2.9, Quality Assurance Records for Nuclear Power Plants, Requirements for Collection, Storage and Maintenance of.

COMMERCIAL KEEPING

Since definite retention times for radiographs are often specified by applicable codes, archival keeping may not always be necessary. Recent studies[‡] have indicated that industrial x-ray films, with a residual thiosulfate ion level of up to 5 micrograms/cm², per side (as measured by the methylene blue method described in ANSI PH4.8),[§] should retain their information for at least 50 years when stored at 0 to 24°C (32 to 75°F) and a relative humidity of 30 to 50 percent. Peak temperatures for short time periods should not exceed 32°C (90°F) and the relative humidity should not exceed 60 percent. Storage conditions in excess of these ranges tend to reduce image stability. The extent of reduced image stability is very difficult to define, due to the great number of conditions that could exist outside of the above suggested storage condition ranges. It should be noted that this does not imply that industrial x-ray films with a total residual thiosulfate content of 5 micrograms/cm², per side, will have archival keeping characteristics. It does, however, suggest that these films will fulfill the needs of most current users of industrial x-ray film requiring a storage life of 50 years or less.

ADDITIONAL STORAGE SUGGESTIONS

Regardless of the length of time a radiograph is to be kept, these suggestions should be followed to provide for maximum stability of the radiographic image:

1. Avoid storage in the presence of chemical fumes.

2. Avoid short-term cycling of temperature and humidity.

3. Place each radiograph in its own folder to prevent possible chemical contamination by the glue used in making the storage envelope (negative preserver). Several radiographs may be stored in a single storage envelope if each is in its own interleaving folder.

4. Never store unprotected radiographs in bright light or sunlight.

5. Avoid pressure damage caused by stacking a large number of radiographs in a single pile, or by forcing more radiographs than can comfortably fit into a single file drawer or shelf.

Other recommendations can be found in ANSI PH1.43.

---

*A term commonly used to describe the keeping quality of x-ray film, defined by the American National Standards Institute as:
"*Archival Storage*—Those storage conditions suitable for the preservation of photographic film having permanent value."
   This term is not defined in years in ANSI documents, but only in residual thiosulfate content (residual fixer) for archival storage.
†These documents may be obtained from the American National Standards Institute, Inc., 1430 Broadway, New York, New York 10018.
‡These studies, called Arrhenius tests, are relatively short-term, elevated-temperature tests conducted under carefully controlled conditions of temperature and humidity which simulate the effects of natural aging.
§The methylene blue and silver densitometric methods produce data as a combination of both sides of double-coated x-ray film.

CHAPTER 10

Fundamentals of Processing

In the processing procedure, the invisible image produced in the film by exposure to x-rays, gamma rays, or light is made visible and permanent. Processing is carried out under subdued light of a color to which the film is relatively insensitive. The film is first immersed in a developer solution which causes the areas exposed to radiation to become dark, the amount of darkening for a given degree of development depending on the degree of exposure. After development, and sometimes after a treatment designed to halt the developer reaction abruptly, the film passes into a fixing bath. The function of the fixer is to dissolve the darkened portions of the sensitive salt. The film is then washed to remove the fixing chemicals and solubilized salts, and finally is dried.

Processing techniques can be divided into two general classes—manual processing (see pages 79 to 85) and automated processing (see pages 85 to 90).

If the volume of work is small, or if time is of relatively little importance, radiographs may be processed by hand. The most common method of manual processing of industrial radiographs is known as the tank method. In this system, the processing solutions and wash water are contained in tanks deep enough for the film to be hung vertically. Thus, the processing solutions have free access to both sides of the film, and both emulsion surfaces are uniformly processed to the same degree. The all-important factor of temperature can be controlled by regulating the temperature of the water in which the processing tanks are immersed.

Where the volume of the work is large or the holding time is important, automated processors are used. These reduce the darkroom manpower required, drastically shorten the interval between completion of the exposure and the availability of a dry radiograph ready for interpretation, and release the material being inspected much faster. Automated processors move films through the various solutions according to a predetermined schedule. Manual work is limited to putting the unprocessed film into the processor or into the film feeder, and removing the processed radiographs from the receiving bin.

GENERAL CONSIDERATIONS

Cleanliness

In handling x-ray films, cleanliness is a prime essential. The processing room, as well as the accessories and equipment, must be kept scrupulously clean and used only for the purposes for which they are intended. Any solutions that are spilled should be wiped up at once; otherwise, on evaporation, the chemicals may get into the air and later settle on film surfaces, causing spots. The thermometer and such accessories as film hangers should be thoroughly washed in clean water immediately after being used, so that processing solutions will not dry on them and possibly cause contamination of solutions or streaked radiographs when used again.

All tanks should be cleaned thoroughly before putting fresh solutions into them (see page 80).

Mixing Processing Solutions

Processing solutions should be mixed according to the directions on the labels; the instructions as to water temperature and order of addition of chemicals should be followed carefully, as should the safe-handling precautions for chemicals given on labels or instruction sheets.

The necessary vessels or pails should be made of AISI Type 316 stainless steel with 2 to 3 percent molybdenum, or of enamelware, glass, plastic, hard rubber, or glazed earthenware. (Metals such as aluminum, galvanized iron, tin, copper, and zinc cause contamination and result in fog in the radiograph.)

Paddles or plunger-type agitators are practical for stirring solutions. They should be made of hard rubber, stainless steel, or some other material that does not absorb or react with processing solutions.

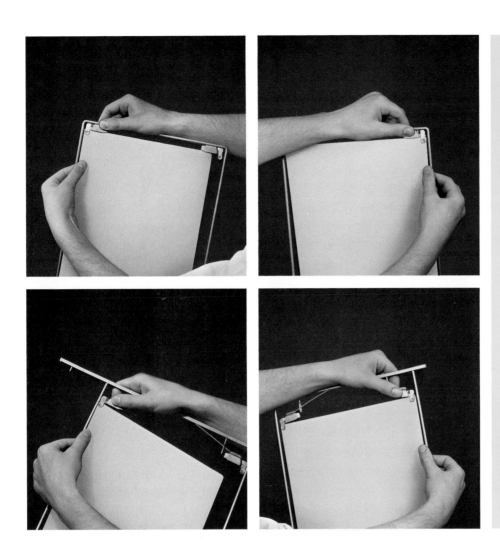

Figure 71—Method of fastening film on a developing hanger. Bottom clips are fastened first, followed by top clips.

Separate paddles or agitators should be provided for the developer and fixer. If the paddles are washed thoroughly and hung up to dry immediately after use the danger of contamination when they are employed again will be virtually nil. A motor-driven stirrer with a stainless steel propeller is a convenient aid in mixing solutions. In any event, the agitation used in mixing processing solutions should be vigorous and complete, but not violent.

MANUAL PROCESSING

When tank processing is used, the routine is, first, to mount the exposed film on a hanger immediately after it is taken from the cassette or film holder, or removed from the factory-sealed envelope. (See Figure 71.) Then the film can be conveniently immersed in the developer solution, stop bath, fixer solution, and wash water for the predetermined intervals, and it is held securely and kept taut throughout the course of the procedure.

At frequent intervals during processing, radiographic films must be agitated. Otherwise, the solution in contact with the emulsion becomes exhausted locally, affecting the rate and evenness of development or fixation.

Another precaution must be observed: The level of the developer solution must be kept constant by adding replenisher. This addition is necessary to replace the solution carried out of the developer tank by the films and hangers, and to keep the activity of the developer constant.

Special precautions are needed in the manual processing of industrial x-ray films in roll form. These are usually processed on the commercially available spiral stainless-steel reels. The space between the turns of film on such a reel is small, and loading must be done carefully lest the turns of film touch one another. The loaded reel should be placed in the developer so that the film is vertical—that is, the plane of the reel itself is horizontal. Agitation in the developer should not be so vigorous as to pull the edges of the film out of the spiral recesses in the reel. The reel must be carefully cleaned with a brush to remove any emulsion or dried chemicals that may collect within the film-retaining grooves.

Cleanliness

Processing tanks should be scrubbed thoroughly and then well rinsed with fresh water before fresh solutions are put into them. In warm weather especially, it is advisable to sterilize the developer tanks occasionally. The growth of fungi can be minimized by filling the tank with an approximately 0.1 percent solution of sodium hypochlorite (Clorox, "101," Sunny Sol bleaches, etc, diluted 1:30), allowing it to stand several hours or overnight, and then thoroughly rinsing the tank. During this procedure, rooms should be well ventilated to avoid corrosion of metal equipment and instruments by the small concentrations of chlorine in the air. Another method is to use a solution of sodium pentachlorphenate, such as Dowicide G fungicide, at a strength of 1 part in 1,000 parts of water. This solution has the advantage that no volatile substance is present and it will not corrode metals. *In preparing the solution, care should be taken to avoid breathing the dust and to avoid contact with the skin, eyes, or clothing.*

Development

DEVELOPER SOLUTIONS Prepared developers that are made ready for use by dissolving in water or by dilution with water provide a carefully compounded formula and uniformity of results. They are comparable in performance and effective life, but the liquid form offers greater convenience in preparation, which may be extremely important in a busy laboratory. Powder chemicals are, however, more economical to buy.

When the exposed film is placed in the developer the solution penetrates the emulsion and begins to transform the exposed silver halide crystals to metallic silver. The longer the development is carried on, the more silver is formed and hence the denser the image becomes.

The rate of development is affected by the temperature of the solution—as the temperature rises, the rate of development increases. Thus, when the developer temperature is low, the reaction is slow, and the development time recommended for the normal temperature would result in underdevelopment. When the temperature is high, the reaction is fast, and the same time would result in overdevelopment. Within certain limits, these changes in the rate of development can be compensated for by increasing or decreasing the time of development.

The time-temperature system of development should be used in all radiographic work. In this system, the developer temperature is always kept within a small range and the time of development is adjusted according to the temperature in such a way that the degree of development remains the same. If this procedure is not carefully observed, the effects of even the most accurate exposure technique will be nullified. Films cannot withstand the effects of errors resulting from guesswork in processing.

In particular, "sight development" should not be used; that is, the development time for a radiograph *should not* be decided by examining the film under safelight illumination at intervals during the course of development. It is extremely difficult to judge from the appearance of a developed but unfixed radiograph what its appearance will be in the dried state. Even though the final radiograph so processed is apparently satisfactory, there is no assurance that development was carried far enough to give the desired degree of film contrast. (See page 141.) Further, "sight development" can easily lead to a high level of fog caused by excessive exposure to safelights during development.

An advantage of standardized time-temperature processing is that by keeping the degree of development constant a definite check on exposure time can always be made. This precludes many errors that might otherwise occur in the production of radiographs. When the processing factors are known to be correct but the radiographs lack density, underexposure can be assumed; when the radiographic image is too dense, overexposure is indicated. The first condition can be corrected by increasing the exposure time; and the second, by decreasing it. The methods for calculating the required changes in exposure are given in detail on pages 53 to 57.

CONTROL OF TEMPERATURE AND TIME Because the temperature of the processing solutions has a decided influence on their activity, careful control of this factor is very important. It should be a rule that the developer be stirred and the temperature be checked immediately before films are immersed in it so that they can be left in the solution for the proper length of time.

Ideally, the temperature of the developer solution should be 68°F (20°C). A temperature below

60°F (16°C) retards the action of the chemical and is likely to result in underdevelopment, whereas an excessively high temperature not only may destroy the photographic quality by producing fog but also may soften the emulsion to the extent that it separates from the base.

When, during extended periods, the tap water will not cool the solutions to recommended temperatures, the most effective procedure is to use mechanical refrigeration. Conversely, heating may be required in cold climates. Under no circumstances should ice be placed directly in processing solutions to reduce their temperature because, on melting, the water will dilute them and possibly cause contamination.

Because of the direct relation between temperature and time, both are of equal importance in a standardized processing procedure. So, after the temperature of the developer solution has been determined, films should be left in the solution for the exact time that is required. Guesswork should not be tolerated. Instead, when the films are placed in the solution, a timer should be set so that an alarm will sound at the end of the time.

AGITATION It is essential to secure uniformity of development over the whole area of the film. This is achieved by agitating the film during the course of development.

If a radiographic film is placed in a developer solution and allowed to develop without any movement, there is a tendency for each area of the film to affect the development of the areas immediately below it. This is because the reaction products of development have a higher specific gravity than the developer and, as these products diffuse out of the emulsion layer, they flow downward over the film surface and retard the development of the areas over which they pass. The greater the film density from which the reaction products flow, the greater is the restraining action upon the development of the lower portions of the film. Thus, large lateral variations in film density will cause uneven development in the areas below, and this may show up in the form of streaks. Figure 72 illustrates the phenomena that occur when a film having small areas whose densities are widely different from their surroundings is developed without agitation of film or developer.

Agitation of the film during development brings fresh developer to the surface of the film and prevents uneven development. In small installations, where few films are processed, agitation is most easily done by hand. Immediately after the hangers are lowered smoothly and carefully into the devel-

oper, the upper bars of the hangers should be tapped sharply two or three times on the upper edge of the tank to dislodge any bubbles clinging to the emulsion. Thereafter, films should be agitated periodically throughout the development.

Acceptable agitation results if the films are shaken vertically and horizontally and moved from side to side in the tank for a few seconds every minute during the course of the development. More satisfactory renewal of developer at the surface of the film is obtained by lifting the film clear of the developer, allowing it to drain from one corner for 2 or 3 seconds, reinserting it into the developer, and then repeating the procedure, with drainage from the other lower corner. The whole cycle should be repeated once a minute during the development time.

Another form of agitation suitable for manual processing of sheet films is known as "gaseous burst agitation." It is reasonably economical to install and operate and, because it is automatic, does not require the full-time attention of the processing room operator. Nitrogen, because of its inert chemical nature and low cost, is the best gas to use.

Gaseous burst agitation consists of releasing bursts of gas at controlled intervals through many small holes in a distributor at the bottom of the processing tank. When first released, the bursts impart a sharp displacement pulse, or piston action, to the entire volume of the solution. As the bubbles make their way to the surface, they provide localized agitation around each small bubble. The great number of bubbles, and the random character of their paths to the surface, provide effective agitation at the surfaces of films hanging in

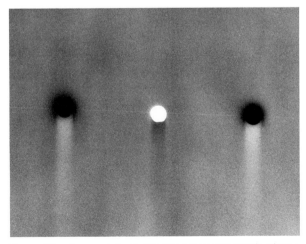

Figure 72—An example of streaking that can result when a film has been allowed to remain in the solution without agitation during the entire development period.

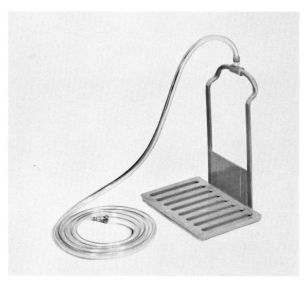

Figure 73—Distribution manifold for gaseous burst agitation.

the solution. (See Figure 73.)

If the gas were released continuously, rather than in bursts, constant flow patterns would be set up from the bottom to the top of the tank and cause uneven development. These flow patterns are not encountered, however, when the gas is introduced in short bursts, with an interval between bursts to allow the solution to settle down.

Note that the standard sizes of x-ray developing tanks will probably not be suitable for gaseous burst agitation. Not only does the distributor at the bottom of the tank occupy some space, but also the tank must extend considerably above the surface of the still developer to contain the froth that results when a burst of bubbles reaches the surface. It is therefore probable that special tanks will have to be provided if the system is adopted.

Agitation of the developer by means of stirrers or circulating pumps should be discouraged. In any tank containing loaded film hangers, it is almost impossible to prevent the uniform flow of developer along certain paths. Such steady flow conditions may sometimes cause more uneven development than no agitation at all.

ACTIVITY OF DEVELOPER SOLUTIONS As a developer is used, its developing power decreases, partly because of the consumption of the developing agent in changing the exposed silver bromide to metallic silver, and also because of the restraining effect of the accumulated reaction products of the development. The extent of this decrease in activity will depend on the number of films processed and their average density. Even when the developer is not used, the activity may decrease slowly because of aerial oxidation of the developing agent.

Some compensation must be made for the decrease in developing power if uniform radiographic results are to be obtained over a period of time. The best way to do this is to use the replenisher system, in which the activity of the solution is not allowed to diminish but rather is maintained by suitable chemical replenishment.

In reference to the *replenisher method* or *replenishment,* the following should be understood. As used here, replenishment means the addition of a stronger-than-original solution, to revive or restore the developer to its approximate original strength. Thus, the replenisher performs the double function of maintaining both the liquid level in the developing tank and the activity of the solution. Merely adding original-strength developer would not produce the desired regenerating effect; development time would have to be progressively increased to achieve a constant degree of development.

The quantity of replenisher required to maintain the properties of the developer will depend upon the average density of the radiographs processed. It is obvious that if 90 percent of the silver in the emulsion is developed, giving a dense image over the entire film, more developing agent will be consumed. Therefore, the developer will be exhausted to a greater degree than if the film were developed to a low density. The quantity of replenisher required, therefore, depends upon the type of subject radiographed. In the processing of industrial radiographs that have a relatively large proportion of dense background, some of the original developer must be discarded each time replenisher is added. The exact quantity of replenisher can be determined only by trial and by frequent testing of the developer.

The replenisher should be added at frequent intervals and in sufficient quantity to maintain the activity reasonably constant for the types of radiographs processed. It is obvious that if replenisher is added only occasionally, there will be a large increase in density of the film after replenishing. By replenishing frequently, these density increases after replenishing are kept at a minimum. The quantity of the replenisher added each time preferably should not exceed 2 or 3 percent of the total volume of the developer in the tank.

It is not practical to continue replenishment indefinitely, and the solution should be discarded when the replenisher used equals two to three times the original quantity of the developer. In any case, the solution should be discarded after three months because of aerial oxidation, and the buildup of gelatin, sludge, and solid impurities.

Arresting Development

After development is complete, developer remaining in the emulsion must be deactivated by an acid stop bath or, if this is not feasible, by prolonged rinsing in clean running water.

If this step is omitted, development continues for the first minute or so of fixation and, unless the film is agitated almost continuously during this period, uneven development will occur, resulting in streakiness.

In addition, if films are transferred to the fixer solution without the use of an acid stop bath or thorough rinsing, the alkali from the developer solution retained by the gelatin neutralizes some of the acid in the fixer solution. After a certain quantity of acid has been neutralized, the chemical balance of the fixer solution is upset and its usefulness is greatly impaired—the hardening action is destroyed and stains are likely to be produced in the radiographs. Removal of as much of the developer solution as possible before fixation prolongs the life of the fixer solution and assures the routine production of radiographs of better quality.

STOP BATH A stop bath consisting of 16 ounces of 28 percent acetic acid per gallon of bath (125 mL per litre), may be used. If the stop bath is made from glacial acetic acid, the proportions should be 4½ ounces of glacial acetic acid per gallon of bath, or 35 mL per litre.
(CAUTION: Glacial acetic acid should be handled only under adequate ventilation, and great care should be taken to avoid injury to the skin or damage to clothing. Always add the glacial acetic acid to the water slowly, stirring constantly, and never water to acid; otherwise, the solution may boil and spatter acid on hands and face, causing severe burns.)

When development is complete, the films are removed from the developer, allowed to drain 1 or 2 seconds (not back into the developer tank), and immersed in the stop bath. The developer draining from the films should be kept out of the stop bath. Instead of draining, a few seconds' rinse in fresh running water may be used prior to inserting the films in the stop bath. This will materially prolong the life of the bath.

Films should be immersed in the stop bath for 30 to 60 seconds (ideally, at 65 to 70°F or 18 to 21°C) with moderate agitation and then transferred to the fixing bath. Five gallons of stop bath will treat about 100 14 x 17-inch films, or equivalent. If a developer containing sodium carbonate is used, the stop bath temperature must be maintained between 65 and 70°F; otherwise, blisters containing carbon dioxide may be formed in the emulsion by action of the stop bath.

RINSING If a stop bath cannot be used, a rinse in *running* water for at least 2 minutes should be used. It is important that the water be running and that it be free of silver or fixer chemicals. The tank that is used for the final washing after fixation should *not* be used for this rinse.

If the flow of water in the rinse tanks is only moderate, it is desirable to agitate the films carefully, especially when they are first immersed. Otherwise, development will be uneven, and there will be streaks in areas that received a uniform exposure.

Fixing

The purpose of fixing is to remove all of the undeveloped silver salt of the emulsion, leaving the developed silver as a permanent image. The fixer has another important function—hardening the gelatin so that the film will withstand subsequent drying with warm air. The interval between placing the film in the fixer solution and the disappearance of the original diffuse yellow milkiness is known as the *clearing time*. It is during this time that the fixer is dissolving the undeveloped silver halide. However, additional time is required for the dissolved silver salt to diffuse out of the emulsion and for the gelatin to be hardened adequately. Thus, the total *fixing time* should be appreciably greater than the clearing time. The fixing time in a relatively fresh fixing bath should, in general, not exceed 15 minutes; otherwise, some loss of low densities may occur. The films should be agitated vigorously when first placed in the fixer and at least every 2 minutes thereafter during the course of fixation to assure uniform action of the chemicals.

During use, the fixer solution accumulates soluble silver salts which gradually inhibit its ability to dissolve the unexposed silver halide from the emulsion. In addition, the fixer solution becomes diluted by rinse water or stop bath carried over by the film. As a result, the rate of fixing decreases, and the hardening action is impaired. The dilution can be reduced by thorough draining of films before immersion in the fixer and, if desired, the fixing ability can be restored by replenishment of the fixer solution.

The usefulness of a fixer solution is ended when it has lost its acidity or when clearing requires an unusually long interval. The use of an exhausted solution should always be avoided because abnormal swelling of the emulsion often results from deficient hardening and drying is unduly prolonged; at high temperatures reticulation or sloughing away of the emulsion may take place. In addition, neutralization of the acid in the fixer solution frequently causes colored stains to appear on the processed radiographs.

Washing

X-ray films should be washed in running water so circulated that the entire emulsion area receives frequent changes. For a proper washing, the bar of the hanger and the top clips should always be covered completely by the running water, as illustrated in Figure 74.

Efficient washing of the film depends both on a sufficient flow of water to carry the fixer away rapidly, and on adequate time to allow the fixer to diffuse from the film. Washing time at 60 to 80°F (15.5 to 26.5°C) with a rate of water flow of four renewals per hour is 30 minutes.

The films should be placed in the wash tank near the outlet end. Thus, the films most heavily laden with fixer are first washed in water that is somewhat contaminated with fixer from the films previously put in the wash tank. As more films are put in the wash tank, those already partially washed are moved toward the inlet, so that the final part of the washing of each film is done in fresh, uncontaminated water.

The tank should be large enough to wash films as rapidly as they can be passed through the other solutions. Any excess capacity is wasteful of water or, with the same flow as in a smaller tank, diminishes the effectiveness with which fixer is removed from the film emulsion. Insufficient capacity, on the other hand, encourages insufficient washing, leading to later discoloration or fading of the image.

The "cascade method" of washing is the most economical of water and results in better washing in the same length of time. In this method, the washing compartment is divided into two sections. The films are taken from the fixer solution and first placed in Section A. (See Figure 75.) After they have been partially washed, they are moved to Section B, leaving Section A ready to receive more films from the fixer. Thus, films heavily laden with fixer are washed in somewhat contaminated water, and washing of the partially washed films is completed in fresh water.

Washing efficiency decreases rapidly as temperature decreases and is very low at temperatures below 60°F. On the other hand, in warm weather, it is especially important to remove films from the tank as soon as washing is completed, because gelatin has a natural tendency to soften considerably with prolonged washing in water above 68°F. Therefore, if possible, the temperature of the wash water should be maintained between 65 and 70°F.

Formation of a cloud of minute bubbles on the surfaces of the film in the wash tank sometimes occurs. These bubbles interfere with washing of the areas of emulsion beneath them, and can subsequently cause a discoloration or a mottled appearance of the radiograph. When this trouble is encountered, the films should be removed from the wash water and the emulsion surfaces wiped with a soft cellulose sponge at least twice during the washing period to remove the bubbles. Vigorous tapping of the top bar of the hanger against the top of the tank rarely is sufficient to remove the bubbles.

Figure 74—Water should flow over the tops of the hangers in the washing compartment. This avoids streaking due to contamination of the developer when hangers are used over again.

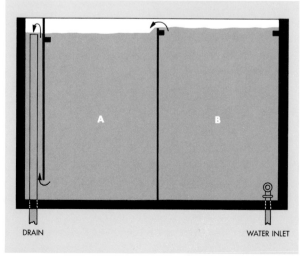

Figure 75—Schematic diagram of a cascade washing unit.

Prevention of Water Spots

When films are removed from the wash tanks, small drops of water cling to the surfaces of the emulsions. If the films are dried rapidly, the areas under the drops dry more slowly than the surrounding areas. This uneven drying causes distortion of the gelatin, changing the density of the silver image, and results in spots which are frequently visible and troublesome in the finished radiograph.

Such "water spots" can be largely prevented by immersing the washed films for 1 or 2 minutes in a wetting agent, then allowing the bulk of the water to drain off before the films are placed in the drying cabinet. This solution causes the surplus water to drain off the film more evenly, reducing the number of clinging drops. This reduces the drying time and lessens the number of water spots occurring on the finished radiographs.

Drying

Convenient racks are available commercially for holding hangers during drying when only a small number of films are processed daily. When the racks are placed high on the wall, the films can be suspended by inserting the crossbars of the processing hangers in the holes provided. This obviates the danger of striking the radiographs while they are wet, or spattering water on the drying surfaces, which would cause spots on them. Radiographs dry best in warm, dry air that is changing constantly.

When a considerable number of films are to be processed, suitable driers with built-in fans, filters, and heaters or desiccants are commercially available.

Marks in Radiographs

Defects, spots, and marks of many kinds may occur if the preceding general rules for manual processing are not carefully followed. Perhaps the most common processing defect is streakiness and mottle in areas which receive a uniform exposure. This unevenness may be a result of:

1. Failure to agitate the films sufficiently during development or the presence of too many hangers in the tank, resulting in inadequate space between neighboring films.
2. Insufficient rinsing in water or failure to agitate the films sufficiently before fixation.
3. The use of an exhausted stop bath or failure to agitate the film properly in the stop bath.
4. In the absence of satisfactory rinsing—insufficient agitation of the films on first immersing them in the fixing bath.

Other characteristic marks are dark spots caused by the spattering of developer solution, static electric discharges, and finger marks; and dark streaks occurring when the developer-saturated film is inspected for a prolonged time before a safelight lamp. If possible, films should never be examined at length until they are dry.

A further trouble is fog—that is, development of silver halide grains other than those affected by radiation during exposure. It is a great source of annoyance and may be caused by accidental exposure to light, x-rays, or radioactive substances; contaminated developer solution; development at too high a temperature; or storing films under improper storage conditions (see page 76) or beyond the expiration dates stamped on the cartons.

Accidental exposure of the film to x-radiation or gamma radiation is a common occurrence because of insufficient protection from high-voltage tubes or stored radioisotopes; films have been fogged through 1/8 inch of lead in rooms 50 feet or more away from an x-ray machine.

AUTOMATED FILM PROCESSING

Automated processing requires a processor (see Figure 76), specially formulated chemicals and compatible film, all three of which must work together to produce high-quality radiographs. This section describes how these three components work together.

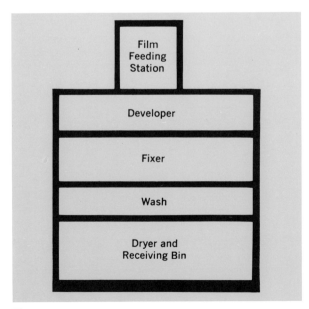

Film Feeding Station
Developer
Fixer
Wash
Dryer and Receiving Bin

Figure 76—An automated processor has three main sections: a film-feeding section; a film-processing section (developer, fixer, and wash); and a film-drying section.

Processing Control

The essence of automated processing is *control,* both chemical and mechanical. In order to develop, fix, wash, and dry a radiograph in the short time available in an automated processor, specifically formulated chemicals are used. The processor maintains the chemical solutions at the proper temperatures, agitates and replenishes the solutions automatically, and transports the films mechanically at a carefully controlled speed throughout the processing cycle. Film characteristics must be compatible with processing conditions, shortened processing times and the mechanical transport system. From the time a film is fed into the processor until the dry radiograph is delivered, chemicals, mechanics, and film must work together.

Automated Processor Systems

Automated processors incorporate a number of systems which transport, process, and dry the film and replenish and recirculate the processing solutions. A knowledge of these systems and how they work together will help in understanding and using automated processing equipment.

TRANSPORT SYSTEM The function of the transport system (see Figure 77) is to move film through the developer and fixer solutions and through the washing and drying sections, holding the film in each stage of the processing cycle for exactly the right length of time, and finally to deliver the ready-to-read radiograph.

In most automated processors now in use, the film is transported by a system of rollers driven by a constant speed motor. The rollers are arranged in a number of assemblies—entrance roller assembly, racks, turnarounds (which reverse direction of film travel within a tank), crossovers (which transfer films from one tank to another), and a squeegee assembly (which removes surface water after the washing cycle). The number and specific design of the assemblies may vary from one model of processor to another, but the basic design is the same.

It is important to realize that the film travels at a constant speed in a processor, but that the speed in one model may differ from that in another. Processing cycles—the time interval from the insertion of an unprocessed film to the delivery of a dry radiograph—in general range downward from 15 minutes. Because one stage of the cycle may have to be longer than another, the racks may vary in size—the longer the assembly, the longer the film takes to pass through a particular stage of processing.

Although the primary function of the transport

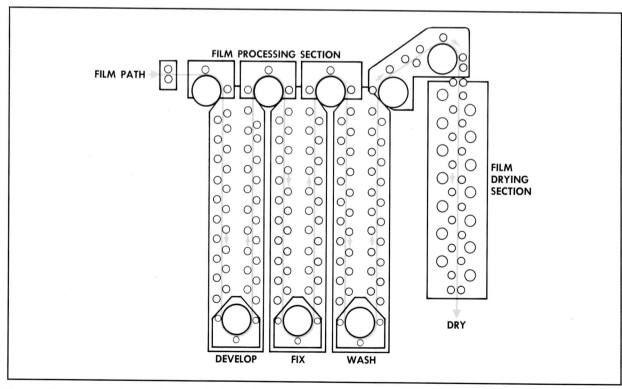

Figure 77—The roller transport system is the backbone of an automated processor. The arrangement and number of its components vary but the basic plan is virtually the same.

system is to move the film through the processor in a precisely controlled time, the system performs two other functions of importance to the rapid production of high-quality radiographs. First, the rollers produce vigorous uniform agitation of the solutions at the surfaces of the film, contributing significantly to the uniformity of processing. Second, the top wet rollers in the racks and the rollers in the crossover assemblies effectively remove the solutions from the surfaces of the film, reducing the amount of solution carried over from one tank to the next and thus prolonging the life of the fixer and increasing the efficiency of washing. Most of the wash water clinging to the surface of the film is removed by the squeegee rollers, making it possible to dry the processed film uniformly and rapidly, without blemishes.

WATER SYSTEM The water system of automated processors has two functions—to wash the films and to help stabilize the temperature of the processing solutions. Hot and cold water are blended to the proper temperature and the tempered water then passes through a flow regulator which provides a constant rate of flow. Depending upon the processor, part or all of the water is used to help control the temperature of the developer. In some processors, the water also helps to regulate the temperature of the fixer. The water then passes to the wash tank where it flows through and over the wash rack. It then flows over a weir (dam) at the top of the tank and into the drain.

Sometimes the temperature of the cold water supply may be higher than required by the processor. In this situation, it is necessary to cool the water before piping it to the processor.

This is the basic pattern of the water system of automated processors; the details of the system may vary slightly, however.

RECIRCULATION SYSTEMS Recirculation of the fixer and developer solutions performs the triple functions of uniformly mixing the processing and replenisher solutions, maintaining them at constant temperatures, and keeping thoroughly mixed and agitated solutions in contact with the film.

The solutions are pumped from the processor tanks, passed through devices to regulate temperature, and returned to the tanks under pressure. This pressure forces the solutions upward and downward, inside, and around the transport system assemblies. As a result of the vigorous flow in the processing tanks, the solutions are thoroughly mixed and agitated and the films moving through the tanks are constantly bathed in fresh solutions.

REPLENISHMENT SYSTEMS Accurate replenishment of the developer and fixer solutions is even more important in automated processing than in manual processing. In both techniques, accurate replenishment is essential to proper processing of the film and to long life of the processing solutions; but, if the solutions are not properly replenished in an automated processor, the film may swell too much and become slippery, with the result that it might get stuck in the processor.

When a film is fed into the processor, pumps are activated which pump replenisher from storage tanks to the processing tanks. As soon as the film has passed the entrance assembly, the pumps stop—replenisher is added only during the time required for a sheet of film to pass through the entrance assembly. The amount of replenisher added is thus related to the size of the sheet of film. The newly added replenisher is blended with the processor solutions by the recirculation pumps. Excess processing solutions flow over a weir at the top of the tanks into the drain.

Different types of x-ray films require different quantities of processing chemicals. It is, therefore, important that the solutions be replenished at the rate proper for the type or types of film being processed, and the average density of the radiographs.

Replenishment rates must be measured accurately and checked periodically. Overreplenishment of the developer is likely to result in lower contrast; slight underreplenishment results in gain of speed and contrast, but severe underreplenishment results in a loss of both. Severe underreplenishment of developer can cause not only loss of density and contrast but also failure of the film to transport at any point in the transport system. Overreplenishment of the fixer does not affect good operation, but is wasteful. However, underreplenishment results in poor fixation, insufficient hardening, inadequate washing, and possible failure of the film to be transported in the fixer rack or at any point beyond.

DRYER SYSTEM Rapid drying of the processed radiograph depends on proper conditioning of the film in the processing solutions, effective removal of surface moisture by the squeegee rollers, and a good supply of warm air striking both surfaces of the radiograph.

Heated air is supplied to the dryer section by a blower. Part of the air is recirculated; the rest is vented to prevent buildup of excessive humidity in the dryer. Fresh air is drawn into the system to replace that which is vented.

Rapid Access to Processed Radiographs

Approximately twelve or fourteen minutes after exposed films are fed into the unit they emerge processed, washed, dried and ready for interpretation. Conservatively, these operations take approximately 1 hour in hand processing. Thus, with a saving of at least 45 minutes in processing time, the holding time for parts being radiographed is greatly reduced. It follows that more work can be scheduled for a given period because of the speed of processing and the consequent reduction in space required for holding materials until the radiographs are ready for checking.

Uniformity of Radiographs

Automated processing is very closely controlled time-temperature processing. This, combined with accurate automatic replenishment of solutions, produces day-after-day uniformity of radiographs rarely achieved in hand processing. It permits the setting up of exposure techniques which can be used with the knowledge that the films will receive optimum processing and be free from processing artifacts. Processing variables are virtually eliminated.

Small Space Requirements

Automated processors require only about 10 square feet of floor space. The size of the processing room can be reduced because hand tanks and drying facilities are not needed. A film loading and unloading bench, film storage facilities, plus a small open area in front of the processor feed tray are all the space required. The processor, in effect, releases valuable floor space for other plant activities. If the work load increases to a point where more processors are needed, they can be added with minimal additional space requirements. Many plants with widely separated exposure areas have found that dispersed processing facilities using two or more processors greatly increase the efficiency of operations.

Chemistry of Automated Processing

Automated processing is not just a mechanization of hand processing, but a system depending on the interrelation of mechanics, chemicals, and film. A special chemical system is therefore required to meet the particular need of automated processing.

When, in manual processing, a sheet of x-ray film is immersed in developer solution, the exposed silver halide grains are converted to metallic silver, but, at the same time, the emulsion layer swells and softens. The fixer solution removes the underdeveloped silver halide grains and shrinks and hardens the emulsion layer. Washing removes the last traces of processing chemicals and swells the film slightly. Drying further hardens and shrinks the emulsion. Therefore, the emulsion changes in thickness and in hardness as the film is moved from one step to the next in processing. In manual processing, these variations are of no importance because the films are supported independently and do not come in contact with other films or any other surfaces.

Automated processing, however, places an additional set of demands on the processing chemicals. Besides developing and fixing the image very quickly, the processing chemicals must prevent the emulsion from swelling or becoming either slippery, soft or sticky. Further, they must prepare the processed film to be washed and dried rapidly.

In automated processors, if a film becomes slippery, it could slow down in the transport system, so that films following it could catch up and overlap. Or it might become too sticky to pass some point and get stuck or even wrap around a roller. If the emulsion becomes too soft it could be damaged by the rollers. These occurrences, of course, cannot be tolerated. Therefore, processing solutions used in automated processors must be formulated to control, within narrow limits, the physical properties of the film. Consequently, the mixing instructions with these chemicals must be followed exactly.

This control is accomplished by hardener in the developer and additional hardener in the fixer to hold the thickness and tackiness of the emulsion within the limits required for reliable transport, as well as for rapid washing and drying.

It is also desirable that automated processing provide rapid access to a finished radiograph. This is achieved in part by the composition of the processing solutions and in part by using them at temperatures higher than those suitable for manual processing of film.

The hardening developer develops the film very rapidly at its normal operating temperature. Moreover, the formulation of the solution is carefully balanced so that optimum development is achieved in exactly the time required for the hardener to harden the emulsion. If too much hardener is in a solution, the emulsion hardens too quickly for the developer to penetrate sufficiently, and underdevelopment results. If too little hardener is in the solution, the hardening process is slowed, overdevelopment of film occurs, and transport problems may be encountered. To maintain the proper balance it is essential that developer solution be replenished at the rate proper for the type or types of film being processed, and the average density of the radiographs.

Because washing, drying, and keeping properties of the radiograph are closely tied to the effectiveness of the fixation process, special fixers are needed for automatic processing. Not only must they act rapidly, but they must maintain the film at the proper degree of hardness for reliable transport. Beyond this, the fixer must be readily removed from the emulsion so that proper washing of the radiograph requires only a short time. A hardening agent added to the fixer solution works with the fixing chemicals to condition the film for washing and for rapid drying without physical damage to the emulsion.

Experience has shown that the solutions in this chemical system have a long life. In general, it is recommended that the processor tanks be emptied and cleaned after 50,000 films of mixed sizes have been processed or at the end of 3 months, whichever is sooner. This may vary somewhat depending on local use and conditions; but, in general, this schedule will give very satisfactory results.

Film-Feeding Procedures

SHEET FILM Figure 78 shows the proper film-feeding procedures. The arrows indicate the direction in which films are fed into the processor. Wherever possible, it is advisable to feed all narrower films side by side so as to avoid overreplenishment of the solutions. This will aid in balanced replenishment and will result in maximum economy of the solutions used.

Care should be taken that films are fed into the processor square with the edge of a side guide of the feed tray, and that multiple films are started at the same time.

In no event should films less than 7 inches long be fed into the processor.

ROLL FILM Roll films in widths of 16 mm to 17 inches and long strips of film may be processed in a *Kodak* Industrial *X-Omat* Processor. This requires a somewhat different procedure than is used when feeding sheet film. Roll film in narrow widths and many strips have an inherent curl because they are wound on spools. Because of this curl, it is undesirable to feed roll or strip film into the processor without attaching a sheet of leader film to the lead-

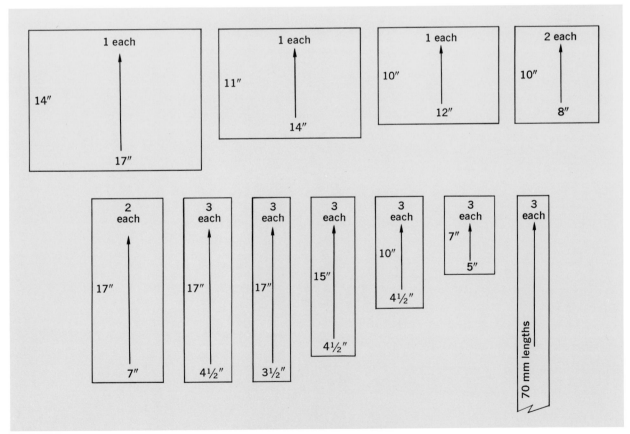

Figure 78—Film-feeding procedures for KODAK Industrial X-OMAT Processors.

ing edge of the roll or strip. Ideally, the leader should be unprocessed radiographic film. Sheet film which has been spoiled in exposure or accidentally light-fogged can be preserved and used for this purpose.

The leader film should be at least as wide as, and preferably wider than, the roll film and be a minimum of 10 inches long. It is attached to the roll film with a butt joint using pressure-sensitive polyester tape, such as *Scotch* Brand Electrical Tape No. 3, one inch in width. (Other types of tape may not be suitable due to the solubility of their bases in the processing solutions.) Care should be taken that none of the adhesive side of the tape is exposed to the processing solutions. Otherwise, the tape may stick to the processor rollers or bits of adhesive may be transferred to the rollers, resulting in processing difficulties.

If narrow widths of roll or strip films are being fed, they should be kept as close as possible to one side guide of the feed tray. This will permit the feeding of standard-size sheet films at the same time. Where quantities of roll and strip films are fed, the replenisher pump should be turned off for a portion of the time. This will prevent overreplenishment and possible upset of the chemical balance in the processor tanks.

FILING RADIOGRAPHS

After the radiograph is dry it must be prepared for filing. With a manually processed radiograph, the first step is the elimination of the sharp projections that are caused by the film-hanger clips. Use of film corner cutters will enhance the appearance of the radiograph, preclude its scratching others with which it may come in contact, facilitate its insertion into an envelope, and conserve filing space.

The radiograph should be placed in a heavy manila envelope of the proper size, and all of the essential identification data should be written on the envelope so that it can be easily handled and filed. Envelopes having an edge seam, rather than a center seam, and joined with a nonhygroscopic adhesive are to be preferred, since occasional staining and fading of the image is caused by certain adhesives used in the manufacture of envelopes.* Ideally, radiographs should be stored at a relative humidity of 30 to 50 percent.

---

*See "American Standard Requirement for Photographic Filing Enclosures for Storing Processed Photographic Films, Plates and Papers," PH1.53-1978, or latest revision thereof. Available from American National Standards Institute Inc., 1430 Broadway, New York, N.Y. 10018.

Process Control

Users of industrial radiography must frequently meet requirements of density and sensitivity set forth in inspection specifications. They must also keep the radiograph rejection rate to a minimum. Control of variability in both exposing and film processing is essential if these requirements are to be met. Although exposure and processing are the most frequent sources of significant variations, other factors, such as intensifying screens and film, also contribute to radiographic variability.

The exposure of industrial x-ray film to energy in the visible spectrum is *not* a reliable indicator of the process level or the repeatability of exposures to x-radiation. Therefore, exposure of the control film to white light is *not* a satisfactory tool for controlling process variability in industrial radiography. In the procedure for controlling either the exposing unit or the film processing, or both, described here, x-radiation is used to expose the control film.

The data obtained from exposures to x-radiation can be utilized in many ways. The procedure is simple and not only reveals both exposing and processing variations but also differentiates between them. It can be customized to fit specific requirements; a few suggestions for doing so are presented later. No attempt has been made to list all the measures available for reducing variations in density inasmuch as the procedures required for adequate process control depend on the conditions in the laboratory and production specifications.

EQUIPMENT AND MATERIALS

Most of the equipment and the material needed to set up a meaningful program of process control is readily available to industrial radiographers. The key items are:

ELECTRONIC DIRECT-READING DENSITOMETER Accurate, precise densitometers capable of measuring diffuse density are available from dealers in photographic supplies.

CALIBRATED FILM STRIP The strip is used to check the precision of the densitometer. If one is not available, a control strip on which previous readings have been recorded will suffice.

STEPPED WEDGE The wedge should be made of steel or the material most often tested. One step should be thick enough to permit the passage of radiation sufficient to produce a density of 0.6 to 1.0; another step should be thin enough to permit passage of twice as much radiation to produce a density of 2.0 or higher. Although not essential, it is helpful if the difference in the density produced by adjacent steps is uniform.

FILM A box of industrial x-ray film of the type most frequently used in the normal production operation should be reserved for the control program. Before all the film in this box has been used, a new box should be reserved for the same purpose and the necessary data on the response of the new film should be obtained.

REFRIGERATOR After exposure, control strips must be stored at 40°F (4.5°C) or lower.* When the stock of control film exceeds a six-week supply, the unexposed film should be stored in a refrigerator and no more than a one-week supply should be removed at any one time. When the temperature and the humidity are high (more than 75°F [24°C] and 50 percent relative humidity), unexposed film should be refrigerated regardless of the size of the stock.

---

*When unprocessed film (exposed as well as unexposed) is stored in the refrigerator, it must be protected from moisture, light, and air. Unopened boxes of control film are protected by a moistureproof inner wrap. A one-week supply of control film and exposed but unprocessed control strips can be protected by placing them in an airtight, lighttight container; putting the container in a bag made of moistureproof material, such as polyethylene; and sealing the bag. When film is taken from refrigeration, the moistureproof wrapper must not be removed until sufficient time has elapsed for the film to reach room temperature to prevent moist air from condensing on the cold surface of the film.

CASSETTE One cassette or film holder for each exposing unit in the operation should be reserved for exclusive use in exposing control film in that unit.

INTENSIFYING SCREENS If lead intensifying screens are used in the normal production operation, one set should be reserved for exclusive use in exposing control film.

ELECTRONIC THERMOMETER WITH SUBMERSIBLE STAINLESS STEEL PROBE This is an essential item if two or more automatic processors are operated at a common density aim that has close tolerances. It is also helpful in reducing variations in density attributable to processing in an automatic processor.

GENERAL ASPECTS

The information that follows pertains to the control system as a whole. Specific details of exposing, processing, use of data accumulated, and the like are presented later.

Procedure

Establish a specific exposure technique for each x-ray unit in the control system. A separate technique for each unit is essential because of variability in units of the same design and variations among units of different designs. Each time control film is exposed in a unit, the technique established for that particular unit must be followed exactly.

Routinely check the accuracy and the precision of the densitometer using the calibrated film strip or a control strip on which previous readings have been recorded.

Maintain a separate process control chart for each exposing unit in the control system.

Maintain a separate process control chart for each processor in the control system unless two or more processors are kept operating at a common level.

Expose control film each day in a designated unit. The exposed film is cut into enough strips to provide a minimum of two strips for each processor in the operation. More than the minimum number of strips are cut if possible.

Identify each strip.

Place half of the strips in an airtight, lighttight container; put the container *in a moistureproof bag* (a bag made of polyethylene, for example); and place the package *in the refrigerator* to minimize fading of the latent image. The strips are kept in the refrigerator for processing with freshly exposed strips the following day. (A freshly exposed control strip is always processed with a refrigerated control strip exposed in the same unit the previous day.)

Tape the control strips to a leader if they are to be processed in an automated processor. This will provide better transport. If they are to receive rack-and-tank processing, it may be necessary to make a special strip hanger or to adapt a standard film hanger in order to hold the strips securely. The control strips should always be processed with the high-density end down.

Maintain an accurate, up-to-date log containing all information that could affect process control. The following should be included in the log:

1. Maintenance data and changes resulting from readjustment of an exposing unit, including the supply of electrical current to the unit and any significant changes in line voltage.

2. Maintenance data and changes resulting from readjustment of a processor.

3. Age of the developer replenisher.

4. Replenishment rate of the developer.

5. Age of the developer.

6. Temperature of the developer at the time control strips are being processed.

7. Comments on fixer and wash and their replenishment rates. (Although the condition of the fixer and the wash does not noticeably affect the variability of film densities, the condition of each does have an effect on the physical quality of processed film.)

Process Control Charts

Two steps on the control strip are selected for measurement. One step should have a density of 0.6 to 1.0; the other, 2.0 or higher. A specific area of the step is selected for measurement, and that same area on each of the two steps is measured to obtain a high-density value and a low-density value. From the first day on, four values are obtained—the value of the high-density step and the value of the low-density step on the fresh control strip and the values of the corresponding steps on the control strip exposed the preceding day—always in the same area on each step.

The upper and the lower control limits for the process density aim can be assigned arbitrarily on the basis of acceptable tolerances in the process operation (2.0 ± 0.2, for example). Sometimes it is desirable to calculate more precise control limits, however; and a statistical method, such as the standard deviation of density values with three sigma control limits, can be used to determine the limits. If the standard deviation with three sigma control limits *is* used, 95 percent of all data collected should be within the limits. (The method of calculating the standard deviation with three sigma control limits, which is given in most books on statistical quality control, has been described in relatively simple terms by Mason.*)

When a new box of control film is introduced into the operation, control exposures are made on both the old and the new stock for four days and a temporary process density aim is computed on the basis of the average densities obtained during the four-day period. After 10 days, the process density aim, or the mean density, is recomputed on the 10-day average.

The process density aim is reestablished whenever changes are made within the operation. Such changes as the introduction of a new control film and alterations to the exposing unit, for example, make it necessary to reestablish the aim.

Control limits for variables of the exposing unit and the film processing are wider than if either were monitored individually.

The densitometric data obtained from the control strips can be utilized in several ways. Figure 79 shows densitometric readings for a 10-day period. Some of the data in Figure 79 are plotted on one process control chart (Figure 80) to show variations in exposure and processing; some are plotted on another process control chart (Figure 81) to indicate changes in contrast.

Process Control Technique

Certain measures can be taken to reduce radiographic process variability, but they depend largely on the conditions and the requirements of the individual user. This is particularly true with respect to variations introduced by the exposing unit. Obvious causes of exposure variability, such as fluctuations or changes in line voltage, must first be eliminated, of course. The following may be of help in reducing variations in density attributable to processing of the film in an automated processor.

Use the electronic thermometer with the submersible stainless steel probe to set the temperature of the developer. Use it to check for fluctuations in developer temperature; the thermostat for the developer may allow the temperature to vary by 2°F. Procedures and thermostats that will hold the temperature of the developer to ± 1/5°F are available.

Use a graduate frequently to check and to maintain the developer replenisher rate at that recommended for the average film density in the process.

*Mason, R. D.: *Statistical Techniques in Business and Economics.* Monograph in Irwin Series in Quantitative Analysis for Business. Published by Richard D. Irwin, Inc., Homewood, Illinois, 1970, pp. 116-123, 314-329.

Keep processing of completely exposed film or completely unexposed film to a minimum. Developer that is overreplenished generally causes an increase in film densities; developer that is underreplenished generally causes a decrease in film densities. When the stage of underreplenishment reaches a certain point, the rate at which film densities decrease may become quite rapid.

Never permit the developer replenisher to age beyond its recommended storage life. If the developer replenisher exceeds its storage life, the storage tank should be emptied and rinsed and a fresh solution should be prepared. *Do not* mix fresh developer replenisher with a solution that is near the limit of its recommended storage life.

Do not replenish with developer replenisher that is past its recommended storage life. Do not replenish with oxidized developer replenisher. The results of these practices can be losses in density, a shift in contrast, or both.

Use fresh solutions at the time a process control system is initiated to reduce the possibility of establishing a process density aim base on a process that is abnormal.

When the processed control strips indicate an out-of-control condition, check for an obvious error, such as the temperature of the developer (if a film processing error is indicated), or the technique used to set up the exposing unit (if an exposure error is indicated), or poor densitometry. If an abnormal process *is* indicated, process another set of control strips. (It is for this purpose that more than the minimum of two strips for each processor is advisable.)

When an out-of-control condition does exist that is the result of a film processing variation and not an obvious testing error, the process should be restarted with fresh solutions. Adding chemicals to the developer is frequently unsuccessful, is more time-consuming, and is more expensive in the end than restarting the process. Contamination of the developer (1 mL of fixer in a gallon of developer can be detected in film densities) and underreplenishment of the developer are examples of conditions that indicate the process should be restarted with fresh solutions.

TECHNIQUE
Exposure of Control Film

Load the cassette with a sheet of control film. (If the control film has been refrigerated, be sure to allow enough time for the film to reach room temperature before handling.) If lead intensifying screens are

DENSITOMETRIC DATA FOR INDUSTRIAL X-RAY PROCESS CONTROL

	Mon.	Tues.	Wed.	Thurs.	Fri.	Sat.	Sun.	Mon.	Tues.	Wed.	Thurs.	Fri.
High-density reading	1.95	2.12										
		1.90	1.85									
			2.15	2.00								
				1.70	1.80							
					1.95	—	—	1.85				
								2.05	2.00			
									1.95	1.80		
										1.90	1.95	
											2.05	1.95
												2.00
Low-density reading	0.82	0.91										
		0.71	0.74									
			1.03	0.90								
				0.61	0.64							
					0.80	—	—	0.80				
								1.01	0.97			
									0.92	0.72		
										0.80	0.88	
											0.95	0.75
												0.80
Density difference	1.13	1.19	1.12	1.09	1.15	—	—	1.04	1.03	1.10	1.10	1.20

Figure 79—Densitometric data for process control in industrial radiography accumulated in accordance with the procedure described in text. In each of the two pairs of figures shown for every day except the first and the weekend, the top figure is the reading determined from the control strip exposed one or three days before—the latent-image control strip. The bottom figure is the reading determined from the control strip exposed and processed that day—the fresh-image control strip. In this example, the density difference is the difference between the high-density reading and the low-density reading of the fresh control strip. Latent-image control strips serve equally well for determining density difference.

PROCESS CONTROL CHART—EXPOSURE AND PROCESSING VARIATIONS

		Mon.	Tues.	Wed.	Thurs.	Fri.	Sat.	Sun.	Mon.	Tues.	Wed.	Thurs.	Fri.
Upper control limit	2.15												
Process density aim	1.95												
Lower control limit	1.75												

Figure 80—Control chart for one exposing unit and one processor showing variations in exposure and processing. The control limits are wider than if either exposure or processing were monitored individually. All the high-density readings are plotted, but only the readings from control strips exposed at the same time are connected. The connecting lines represent the day-to-day repeatability of film processing; the difference between point plots on a given day represents the repeatability of exposure.

used in the normal production operation, use the set reserved for process control.

Carefully set up the x-ray exposure unit for the established exposure technique. The cassette and the stepped wedge must be positioned identically each time control film is exposed.

Cut the exposed film into a minimum of twice the number of processors in the control system. If possible, cut more than the minimum, but do *not* make duplicate exposures and assume they are identical. Each exposure of control film must be considered a complete control.

Identify the strips as to date and exposing unit.

Place half of the exposed control strips in an airtight, lighttight, and moistureproof wrapper and store them in the refrigerator. Remove the moistureproof package of control strips exposed in the *same unit* the previous day. When these latent-image control strips are at room temperature, they can be handled and processed.

Processing of Control Strips

Process the freshly exposed control strips and the strips exposed in the same unit the previous day in the corresponding processor (or processors).

If the strips are less than 2 1/2 inches wide, tape them to a leader. If a leader is not used, process a cleanup sheet before processing the control strips.

Always process a fresh control strip exposed in a designated unit with a control strip exposed in that same unit the previous day.

Record in the log the temperature of the developer at the time the strips are in the processor. Record in the log any other information pertinent to process control.

Densitometry

Check the accuracy and the precision of the densitometer each time *before* it is used to obtain the numerical density values from the control strips.

Determine the density value of each of the two steps selected for measuring density on the freshly exposed and processed control strip and the density of the two corresponding steps on the latent-image control strip. (Once these steps have been selected, the density of the same area on each of the steps is used every time control strips are measured.)

Record the densitometric readings and plot them on the process control charts.

DISCUSSION

Densitometric data and process control charts for one exposing unit and one processor are presented as Figures 79 to 81. The process density aim (the mean density) and the upper and lower control limits on the charts (Figures 80 and 81) were computed from the densitometric data for the 10-day period shown in Figure 79. As stated earlier, the control limits on both charts are wider than would be the case if either the exposing unit or the processor were monitored individually. Although two control charts are illustrated, they can be combined into one for convenience.

Exposure and Processing Variations

Variations in both exposure and processing are reflected on the process control chart reproduced as Figure 80. All the high-density readings are plotted, but only the readings from control strips exposed at the same time are connected. The lines represent the day-to-day repeatability of film processing; the difference between the point plots on a given day represents the repeatability of exposure.

There will be a slight density loss of the latent image on the control strips processed one to three days after exposure even though they are refrigerated. However, this small difference in density is no cause for concern with the type of control chart shown as Figure 80. If the density of the refrigerated control strip is always slightly lower (0.03 to 0.04) than that of its corresponding fresh control strip, the process is repeating identically.

Contrast Variations

Variations in film contrast are reflected on the control chart reproduced as Figure 81. The difference between the high-density reading and the low-density reading of the fresh-image control strips is plotted on this chart. (Latent-image control strips serve equally well for determining density difference.) A decrease in density difference from the mean density, or the process density aim, indicates lower contrast; an increase in density difference indicates higher contrast.

PROCESS CONTROL CHART—CHANGES IN FILM CONTRAST

		Mon.	Tues.	Wed.	Thurs.	Fri.	Sat.	Sun.	Mon.	Tues.	Wed.	Thurs.	Fri.
Upper control limit	1.30												
Process density aim	1.10												
Lower control limit	0.90												

Figure 81—Control chart for one exposing unit and one processor indicating changes in contrast. The difference between the high-density reading and the low-density reading of the fresh-image control strips is plotted, and the plots are connected. A decrease in density difference from the process density aim, or the mean density, indicates lower contrast; an increase indicates higher contrast.

The Processing Room

The location, design, and construction of the x-ray processing facilities are major factors in the installation of adequate radiographic services. These facilities may be a single room, or a series of rooms for individual activities, depending on the amount and character of the work performed. Because of the special importance of these rooms for the handling, processing, and storing of x-ray films, both their general and detailed features should be most thoughtfully worked out. When planning reflects care and foresight, the effort expended is soon offset by ease of operation, improved production, and lowered costs of maintenance.

The flow of x-ray films from the radiographic room, through the processing facilities, to the viewing room should be a simple yet smooth operation requiring the fewest possible steps. The routine can be expedited by proper planning of the location within the department of the room or rooms devoted to processing, and by efficient arrangement of the equipment.

Ideally, processing rooms should be supplied with filtered air, at a pressure above that of the outside. This is particularly important when the outside air is likely to be contaminated with sand, dirt, or other airborne particles.

PROCESSING AREA

The volume of films to be handled in the department, and the importance of rapid access to the finished radiographs, will determine whether manual or automatic processing will be used.

Manual Processing

If the work load is small or intermittent, a single room containing all of the facilities can be used (Figure 82). However, if the volume of manual processing is relatively high, production can be expedited by dividing the operations among three areas: A room for loading and unloading cassettes; a processing room with a through-the-wall tank; and a washing and drying room.

In general, the manual processing room should be large enough to hold all the necessary equipment without crowding. However, there is no advantage in having excessive floor space, although need for future expansion should be anticipated. The room shown in Figure 82 will permit the processing of more than 200 films a day, and can be constructed in a floor space 9½ x 15 feet.

It is most efficient to have the processing area adjoin the exposure room. However, in departments where highly penetrating radiation is used, the amount of radiation shielding needed to protect personnel and film may be prohibitively expensive, in which case the processing room must be located at a safe distance.

LOADING BENCH Basically, operations performed in the processing areas should be separated into parts—the "dry" and the "wet." The dry activities—such as the handling of unprocessed film, loading and unloading of cassettes and exposure holders, and the loading of processing hangers—are all done at the loading bench. This may be either opposite the processing tanks in the same room or in a separate adjacent room. Where a cassette-transfer cabinet is used, it should open onto the loading bench, which should provide facilities for storage of processing hangers and other items, and a lighttight film bin. Items such as the transfer cabinet, film storage bin, and processing hanger brackets are commercially available.

PROCESSING TANKS Processing the films, which involves the wet activities of developing, stopping, fixing, and washing, should be carried out in an area separate from the loading bench. This arrangement is designed to avoid splashing solutions on screens, films, and loading areas and, in general, to prevent interference with loading-bench operations.

The tanks must be constructed of a corrosion-resistant material. The majority are now being fab-

LAYOUT PLAN

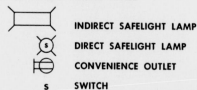

LEGEND

1	LIGHT LOCK WITH LIGHTTIGHT DOORS (SEE ALSO FIGURE 84)	13	AIR EXHAUST FROM FILM DRYER
2	LOADING BENCH	14	X-RAY PROCESSING TANK
3	FILM STORAGE BIN		a. DEVELOPER
4	LIGHTTIGHT DRAWER		b. STOP BATH
5	WASTE BIN		c. FIXER
6	FILM DRYER		d. WASH (CASCADE)
7	CASSETTE AND FILM HOLDER STORAGE		e. SINK
8	PASS BOX	15	ELECTRIC TIMER
9	FILM HANGER RACKS	16	CHART BOARD
10	SUPPLY CABINET	17	ILLUMINATOR
11	AIR SUPPLY DUCT	18	DRAINAGE RACK FOR HANGERS
12	LIGHTTIGHT LOUVRE	19	LIGHTTIGHT ACCESS PANEL

INDIRECT SAFELIGHT LAMP

DIRECT SAFELIGHT LAMP

CONVENIENCE OUTLET

s   SWITCH

X-RAY PROCESSING ROOM

Figure 82—Plan of a manual x-ray processing room.

ricated of AISI Type 316 stainless steel with 2 to 3 percent of molybdenum. Special techniques must be employed in the fabrication of these tanks to avoid corrosion in the welded areas.

The film capacity of the entire processing area is determined by the size of the insert tanks. Based on a 5-minute development time, a 5-gallon developer tank can handle 40 films an hour with four hangers being handled simultaneously, and allowing for the time during which the hangers are removed and inserted in the stop bath. The capacity of the stop bath tank should be equal to that of the developer tank, and the fixer tank should be at least twice as large as the developer tank. The washing tank should hold at least four times the number of hangers accommodated in the developer tank.

FILM DRYERS One of the important considerations in designing the processing area is the film dryer. It should be fast-acting without overheating the film. Hot air, infrared, and desiccant dryers are commercially available. Whenever possible, a filter should be inserted in the air intake. This may, however, create such a resistance to the airflow as to require a fan of larger capacity than would be needed without the filter. A removable drip pan beneath each film compartment or drawer is useful as an aid in keeping the dryer clean. As a precaution, heating elements should be connected in the fan circuit so that heat cannot be turned on without turning on the fan.

Automated Processing

The chief difference between processing rooms for manual and automated processing is the absence of the space-consuming processing tanks. The only part of the automated processor that need be in the processing room is the film-feeding station, and this is quite small. The plans can follow the general form of Figure 83. Note the provision in the outer (light) room for mixing and storing processing chemicals and washing processor components.

In planning a new processing room for an automated processor, early consideration should be given to providing the water, electrical, drainage, and exhaust facilities required by the processor.

GENERAL CONSIDERATIONS

There are a number of considerations that apply to all processing rooms, whether for manual or automatic processing.

Entrances

Three general types of entrances are used for the processing room: The single door, the light lock (double or revolving doors), and the labyrinth or maze. The single door is shown in Figure 83 and a double-door light lock in Figure 82.

Which is best suited to a particular installation is determined largely by the traffic in and out of the processing room and by the amount of floor space available. The single door equipped with an inside bolt or lock is most economical of floor space and is practical where one employee handles the processing. However, in most cases a labyrinth, or a vestibule with two interlocking doors, is generally employed. Plans for the double-door and revolving door light locks, as well as a labyrinth, are shown in Figure 84.

Wall Covering

The walls of the processing room can be of any pleasing color. A cream or buff will give maximum reflectance for safelight illumination. A good semigloss paint is satisfactory for any wall where chemicals are not likely to be spattered. The best protective materials for walls near the processing tanks in a manual processing room or in a chemical mixing area are ceramic tile, structural glass sheets, or stainless steel. Care should be taken in choosing tile since there have been instances when radioactive material has been incorporated in the glaze of the tile. Corrosion- and stain-resistant paints are available but do not have the permanence of stainless steel, tile, or structural glass.

Floor Covering

The ideal floor is resistant to chemical corrosion and staining, of waterproof installation, of a suitable color, and free from slipperiness. Porcelain and the natural clay tiles are satisfactory, as are the darker asphalt tiles. Linoleum, and plastic and rubber tiles are less desirable because they may be stained or pitted by the processing solutions.

Plumbing

In drainage lines, the greatest problem encountered is corrosion. Stainless steel, glass, chemical stoneware, and anticorrosion iron are usually satisfactory. Galvanized steel may be used when waste solutions do not remain in pipes. Under no circumstances should two metals be used, such as copper pipe with galvanized steel fittings, because of the likelihood of corrosive electrolytic action. Plastic fittings will eliminate this problem.

Lines carrying processing or replenisher solutions from storage tanks must be of stainless steel, glass, plastic, or other inert, corrosion-resistant material.

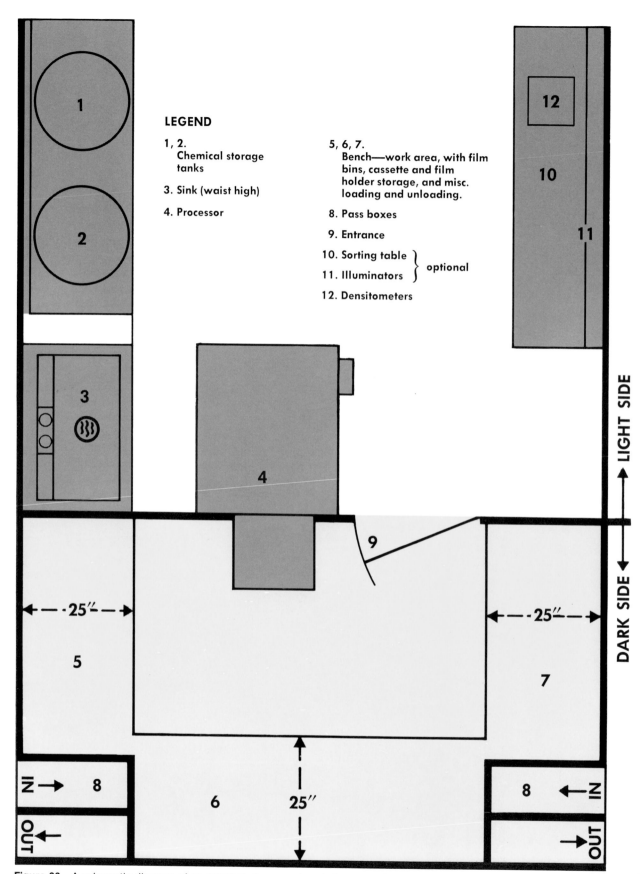

LEGEND

1, 2.
Chemical storage
tanks

3. Sink (waist high)

4. Processor

5, 6, 7.
Bench—work area, with film
bins, cassette and film
holder storage, and misc.
loading and unloading.

8. Pass boxes

9. Entrance

10. Sorting table ⎫
11. Illuminators ⎬ optional
                  ⎭

12. Densitometers

LIGHT SIDE

DARK SIDE

Figure 83—A schematic diagram of an automated processing darkroom and adjacent light-room area.

Illumination

The processing area must be provided with both white light and safelight illumination. White light is desirable for many activities, including cleaning and maintenance.

Since excessive exposure of film to safelight illumination will result in fog, the arrangement of the safelight lamps must be carefully considered. A manual processing room should be divided into three zones of safelight intensity: The brightest, in which the films are washed and placed in the dryer; the medium zone, where films are developed and fixed; and the dimmest zone, where loading-bench activities are carried on. Only one level of illumination is usually provided in a processing room for automatic processing, since the manipulation of unprocessed film is reduced to a minimum.

The "safeness" of processing room illumination depends equally on the use of the proper safelight filter, the use of the proper wattage of bulb, the proper placement of lamps with respect to film, and not exceeding the maximum "safe" time of exposure of the film to safelight illumination.

Exposed films are more sensitive to fogging from the safelight illumination than are unexposed films. Hence, it is especially important to guard the exposed films against prolonged exposure to safelight illumination. Note that the screen-type films are more sensitive to fogging by safelight illumination than direct-exposure films.

A simple method of checking the safety of illumination is to test it with the fastest film used in the laboratory, as follows: An exposure is made of a stepped wedge. In the processing room, the holder is unloaded and the film placed in the area where it is normally handled. Part of the film is covered with opaque paper. The remainder is exposed to the safelight illumination for the maximum time normally needed for handling. The test film is then given standard processing. If no density shows on the uncovered part that received the safelight exposure, as compared with the covered part, the lighting may be assumed to be safe.

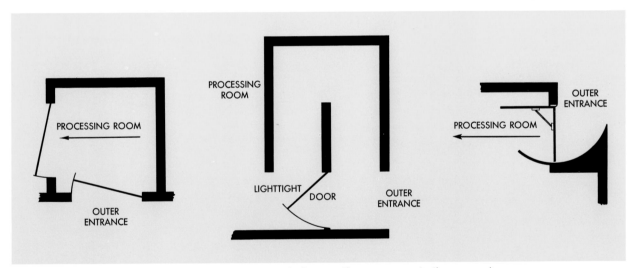

Figure 84—Light locks (left and right) and maze (center) allow continuous access to the processing room.

Special Processing Techniques

There are a number of special techniques useful in the processing of x-ray films. Some are applicable to both manual and automatic processing and others only to manual. Certain radiographic installations may use one or more of these routinely; others may employ them as circumstances warrant.

INTENSIFICATION OF UNDEREXPOSED RADIOGRAPHS

Every industrial radiographic department occasionally encounters a radiograph which has been underexposed, either through an oversight or because of insufficient machine capacity. If the radiograph cannot be repeated because the time required for proper exposure would be prohibitively long, or if the item is no longer available, the underexposed negative can in many cases be salvaged. Chemical intensification of the completely processed film may, under certain circumstances, result in a speed gain of a factor of 5 to 7, while still retaining acceptable radiographic quality.

Films may be intensified immediately after fixation, or after they have been fully washed and dried. In either case, the films are washed in running water for 5 to 10 minutes, hardened in KODAK Special Hardener SH-1 (formula given below) for 5 minutes, and again washed for 5 minutes.

KODAK Special Hardener SH-1

	Avoirdupois, U.S. Liquid	Metric
Water	16 fl oz	500 mL
KODAK Formaldehyde, about 37% solution by weight	2½ fl dr	10.0 mL
KODAK Sodium Carbonate (Monohydrated)	90 grains	6.0 grams
Water to make	32 fl oz	1.0 litre

They are then treated, one at a time, for up to 10 minutes in KODAK Intensifier In-6. The working intensifier is mixed from the stock solutions (formulas opposite) by taking one part of Solution A, and adding in succession two parts of Solution B, two parts of Solution C, and finally one part of Solution A. The order of mixing is important and should be followed. The hardening and intensifi-

cation can conveniently be done in trays. The film should be agitated frequently during intensification, after which it is washed for 20 to 30 minutes in running water and dried normally.

The intensification may be carried out in room light. During treatment, the film may be viewed on an illuminator and the process stopped at any time that the results suit the operator. Intensification in In-6 produces a rather grainy, yellowish image, which is not quite as permanent as a properly fixed and washed silver image. However, sufficient improvement is made in the radiographic sensitivity of underexposed radiographs to make these drawbacks relatively minor.

Because the intensified image is destroyed by acid hypo, under no circumstances should the in-

KODAK Quinone-Thiosulfate Intensifier In-6

	Avoirdupois, U.S. Liquid	Metric
Solution A		
*Water (about 70°F)	96 fl oz	750 mL
†Sulfuric acid (concentrated)	4 fl oz	30.0 mL
KODAK Potassium Dichromate (anhydrous)	3 ounces	22.5 grams
Water to make	1 gallon	1.0 litre
Solution B		
*Water (about 70°F)	96 fl oz	750 mL
KODAK Sodium Bisulfite (anhydrous)	½ ounce	3.8 grams
KODAK Hydroquinone	2 ounces	15.0 grams
KODAK PHOTO-FLO 200 Solution	½ fl oz	3.8 mL
Water to make	1 gallon	1.0 litre
Solution C		
*Water (about 70°F)	96 fl oz	750 mL
KODAK Sodium Thiosulfate (Hypo)	3 ounces	22.5 grams
Water to make	1 gallon	1.0 litre

*The water used for mixing the solutions for the intensifier should not have a chloride content greater than about 15 parts per million (equivalent to about 25 parts of sodium chloride per million); otherwise, the intensification will be impaired. If in doubt as to chloride content, use distilled water.

†**Caution:** Always add the sulfuric acid to the water slowly, stirring constantly, and never the water to the acid; otherwise, the solution may boil and spatter the acid on the hands and face, causing serious burns.

tensified negatives be placed either in a fixing bath or in wash water contaminated with fixing bath. Films to be intensified should be handled as little as possible, and then only by the edges or corners.

The stock solutions from which the intensifier is mixed will keep in stoppered bottles for several months, and the mixed intensifier is stable for 2 to 3 hours. The bath should be used only once and then be discarded because a used bath may produce a silvery scum on the surface of the image.

REMOVAL OF FIXING AGENTS

When, in manual processing, the capacity of the film washing tanks is insufficient, when time must be conserved, or when, as in field radiography, the water supply is limited, the use of KODAK Hypo Clearing Agent between fixation and washing is advantageous. This material permits a reduction of both the time and the amount of water necessary for adequate washing.

After fixation, the excess fixer is removed from the film by a 30-second rinse in water. It is then immersed in KODAK Hypo Clearing Agent solution for 1 to 2 minutes, with agitation. With this procedure, the capacity of the Hypo Clearing Agent bath will be about 750 to 1000 films (8 x 10-inch) or 250 to 330 films (14 x 17-inch) per 5 gallons of solution. If no rinse is used after fixation, the capacity of the bath will be reduced to about 200 to 300 films (8 x 10-inch). The bath should be considered exhausted when that number of films has been processed, or sooner if a precipitation sludge appears. It must then be replaced, not replenished.

After treatment with the Hypo Clearing Agent, films should be washed for 5 minutes, using a water flow which will give a complete change of water 4 to 8 times per hour. However, if water supplies are severely limited, films may be washed in standing water, rather than running water, by soaking for 10 minutes with occasional agitation. The water in the wash tank should be replaced after 10 films (8 x 10-inch) per gallon have been washed.

The effectiveness of the washing procedure and the capacity of the Hypo Clearing Agent bath may be checked by testing a processed film for fixer removal as described in the following section.

TESTING FOR FIXER REMOVAL

Fixing chemicals not adequately removed from films by washing will, over a period of time, cause staining of the film and fading of the developed image. When it is known that films must be pre-

served indefinitely or when there is doubt as to the adequacy of the washing procedures, the amount of fixing chemicals remaining in the film after washing should be determined. This can be done in one of two ways.

ARCHIVAL WASHING Films of archival interest—and this includes the majority of industrial radiographs for code work—should remain unchanged for long periods of time under good storage conditions.* Archival washing for this indefinite preservation of films is defined by American National Standards Institute (ANSI) documents in terms of the concentration of residual thiosulfate in the film. Acceptable methods for measurement are described in ANSI PH4.8-1971,"Methylene-Blue Method for Measuring Thiosulfate, and Silver Densitometric Method for Measuring Residual Chemicals in Films, Plates and Papers."† The methylene-blue method described in this document measures directly the concentration of thiosulfate ion. The silver-densitometric method measures thiosulfate as well as other residual chemicals and requires that a calibration curve be used relating the silver density produced to the thiosulfate content as measured by the methylene-blue method.

For test films or any other films intended for archival keeping, the method for determining residual thiosulfate should be chosen from those covered in the ANSI standard mentioned above. Note that while KODAK Hypo Estimator and KODAK Hypo Test Solution HT-2 (the HT-2 test) provide a quick, convenient means for estimating the amount of hypo (thiosulfate ion) retained in the emulsion, they cannot be used to determine the concentration of residual thiosulfate in terms of archival washing standards.

The methylene-blue method measures *only* thiosulfate. The technique is complex, and the sample must be tested within two weeks of processing. The silver densitometric method measures thiosulfate and other residual chemicals. The technique is simpler, and the results are not affected as much by the length of time between processing and testing. Like the HT-2 test, the silver densitometric method lacks sensitivity at low levels of thiosulfate. It is not sensitive enough to measure thiosulfate reliably below about 0.9 μg per square centimetre. The two procedures for the methylene-blue method described in ANSI PH4.8-1978

---

*American National Standard Practice for Storage of Processed Safety Photographic Film, PH1.43-1979. Published by American National Standards Institute, Inc., New York, New York.
†Available from American National Standards Institute, Inc., 1430 Broadway, New York, New York 10018.

cover the range of 0.1 to 45 μg of thiosulfate ion ($S_2O_3^{--}$) per square centimeter of the test sample. This is the only method ANSI considers sufficiently reliable for determining such a low concentration as 0.7 μg of thiosulfate ion per square centimetre.

METHYLENE-BLUE METHOD Two variations of this method for measuring the concentration of residual thiosulfate are described in detail in ANSI PH4.8-1978. One covers a range of 0.1 to 0.9 μg of thiosulfate ion per square centimetre; the other, a range of 0.9 to 45 μg of thiosulfate ion per square centimetre. If the film is double coated, the residual thiosulfate ion is assumed to be divided equally between the two sides. Therefore, the concentration per square centimetre of emulsion is one-half of the total determined by either variant of the methylene-blue method. The maximum permissible concentration of residual thiosulfate ion for coarse-grain films, such as industrial x-ray films, is 3 μg of thiosulfate or 2 μg of thiosulfate ion per square centimetre.

Either way, testing must be done within two weeks of processing. Both require several reagents, a photometer or a spectrophotometer, and a calibration curve. Tests are conducted as follows:

Residual thiosulfate is extracted from a test sample and reduced to a sulfide. The sulfide reacts with test reagents to form methylene blue. The absorbance or the transmittance of the blue color is then measured with a photometer or a spectrophotometer, and the thiosulfate level is read from a calibration curve.

The methylene-blue method is a complex multistep procedure that requires special materials and equipment and specialized analytic techniques not readily available to most industrial radiographers. Complete step-by-step directions for both procedures, including preparation of the test sample (which must be taken from an area of minimum density—preferably an unexposed but processed area), the various reagents, and the calibration curve, as well as information on the interpretation of results, are given in ANSI PH4.8-1978.

REMOVAL OF ONE EMULSION FROM DOUBLE-COATED FILM

In some applications of x-rays, for example, x-ray diffraction or microradiography, it may be desirable to avoid the parallax associated with an image on double-coated film.

The emulsion can be removed from one side of a processed x-ray film by the following procedure. The film is processed in the normal manner. It need not be dried unless desired. If dry, the film is fastened to a sheet of glass using waterproof tape, and the emulsion surface to be removed is rubbed with a cotton swab saturated with one-half normal potassium hydroxide (28 grams of potassium hydroxide per litre). If wet, it may be pressed firmly to a sheet of glass, and the potassium hydroxide solution applied, care being taken to prevent the solution from flowing onto the glass plate or in any way coming in contact with the bottom emulsion. Care must be exercised to prevent the dry chemical or the solution from coming in contact with the bare skin (use rubber gloves), clothing, or the emulsion surface which is to be preserved. After the film has been rubbed with the swab for about 1 minute, the emulsion is usually soft enough to be scraped off with a smooth, dull implement which will not scratch the film base—for example, a plastic windshield scraper. After the emulsion has been removed, the film is rinsed in running water, removed from the glass plate, and immersed in the fixing bath for a few seconds to neutralize any remaining caustic. It is then washed for about 20 minutes and dried, although a shorter washing period is acceptable if the film is not to be kept indefinitely.

Alternatively, the unwanted emulsion may be covered, prior to development, with some waterproof sheet material which will protect it from the action of the developer. The protective material is removed after development but before fixation. The time of fixation should be extended, since the undeveloped emulsion clears more slowly than does the developed one. Wash and dry according to standard procedure.

If the desired area is narrow, as in the case of a microradiograph or a powder x-ray diffraction pattern, it may be covered with a waterproof adhesive tape. It is inadvisable to overlap strips of tape to cover wider areas because the adhesive coating of the tape does not always adhere tightly enough to the back of the adjoining layer to prevent leakage of the developer. Any tape used should be tested for its impermeability to developer, and to be certain that the adhesive adheres to the tape when the latter is removed rather than to the emulsion.

For larger areas, as in conventional radiography, two films may be taped together, with the unwanted emulsions in contact, and developed together. After development the films are separated, fixed, and washed individually. Alternatively, films may be taped, unwanted side down, to a discarded radiograph or a piece of film base, and removed after development but before fixation.

TRAY PROCESSING

This method is not as efficient as the tank system for the manual processing of x-ray films. However, when tanks are not available, satisfactory results can be obtained by employing trays and exercising the necessary care. The time and temperature recommendations for tank processing apply to tray processing.

Several glass, hard rubber, plastic, or enameled trays are essential. They should be large enough to accommodate the largest film that is used. One tray is used for developer solution, a second for stop bath or rinse water, a third for fixer solution, and a fourth for wash water.

In the tray system, a quantity of solution should be mixed at regular intervals and kept in glass bottles or glazed jugs. Then enough solution to cover the films to a depth of at least 1 inch is poured into the proper trays just before processing.

When the film is removed from the cassette, film holder, etc, a stainless steel clip should be attached to one corner to facilitate handling. The film is then immersed in the developer solution using a quick, sliding motion. If the emulsion is not covered evenly, a dark line will show in the radiograph where the solution pauses. Likewise, dark spots will appear on the film at points where spattered drops of developer strike the dried emulsion. During development, the film should be moved frequently and turned over so that the under side does not adhere to the tray and thereby retard the action of the chemicals. The tray should also be rocked in an irregular manner to provide continual mixing and redistribution of the solution over both surfaces of the film. Similar agitation is necessary in the fixer. Provision should be made for a constant flow of water in the wash tray. Care should be taken to be sure that radiographs do not cling to one another or stick to the bottom of the tray during the course of the washing process.

COMMERCIAL WASHING Films intended for ordinary commercial use should show no image change for several years under normal storage conditions. Adequate washing reduces the residual fixer content of a processed film to an acceptable level. The KODAK Hypo Estimator used with the KODAK Hypo Test Solution HT-2 provides a simple, convenient method for measuring washing efficiency and can be used for cursory estimates of the keeping quality of films. It is especially useful for comparing variations within a test or for comparing several films in the same process. It has the additional advantages of being fast and easy to do.

KODAK Hypo Test Solution HT-2

	Avoirdupois, U.S. Liquid	Metric
Water	24 fl oz	750.0 mL
*Kodak 28% Acetic Acid	4 fl oz	125.0 mL
KODAK Silver Nitrate (Crystals)	¼ ounce	7.5 grams
Water to make	32 fl oz	1.0 litre

*To make approximately 28% acetic acid from glacial acetic acid, dilute 3 parts of glacial acetic acid with 8 parts of water.

Store the solution in a screw-cap or glass-stoppered brown bottle away from strong light. Avoid contact of test solution with the hands, clothing, negatives, prints, or undeveloped photographic materials; otherwise, black stains will ultimately result.

The KODAK Hypo Estimator consists of four color patches reproduced on a strip of transparent plastic. It is used in conjunction with KODAK Hypo Test Solution HT-2. For use in the test, an unexposed piece of film of the same type is processed with the radiographs whose fixer content is to be determined. After the test film is dried, one drop of the KODAK HT-2 Solution is placed on it and allowed to stand for 2 minutes. The excess test solution is then blotted off, and the stain on the film compared with the color patches of the KODAK Hypo Estimator. The comparison should be made on a conventional x-ray illuminator. Direct sunlight should be avoided since it will cause the spot to darken rapidly.

For commercial use, the test spot should be no darker than two thicknesses of Patch 4 of the Hypo Estimator. Two thicknesses can be obtained by folding the estimator along the center of the patch.

STORAGE CONDITIONS The residual fixer concentration for commercial use can generally be tolerated in areas where the average relative humidity and temperature in the storage space are not excessive. These quantities may, however, be excessive when storage conditions are worse than average for temperature and humidity. Archival processing should be the rule whenever it is known that relative humidity and temperature are likely to be constantly excessive, as is the case in tropical and subtropical areas.

SILVER RECOVERY FROM FIXING SOLUTIONS

Silver recovery is both a significant means of conserving a natural resource and a potential source of revenue for users of radiographic products. Kodak silver recovery equipment consists of two simple, nonmoving parts: the KODAK Chemical Recovery

Cartridge, Type 1-P, and the KODAK Circulating Unit, Type P. This equipment is specifically designed for the removal of silver from the overflow streams of automatically replenished processors. It is, however, equally adaptable for use with batch replenishment or hand-processing tanks. The equipment is low in cost, requires only simple nonelectrical installation, and operates at high efficiency (99 percent recovery). There is very little maintenance required.

A single cartridge is sufficient to handle the fixer overflow in almost every instance. The cartridge is connected by means of a flexible plastic tube to the fixer overflow line of a KODAK Industrial X-OMAT Processor or to a storage tank. Simple records of the quantity of fixer and replenisher used are all that need be kept. The exhaustion point of the cartridge is determined by the use of test papers.

The KODAK Chemical Recovery Cartridge is a plastic-lined drum packed with steel wool. The steel wool is gradually dissolved by the acid in the fixer solution. The dissolved iron then replaces the silver in the silver-hypo complex. Since silver is not soluble in the acid in the fixer, it precipitates and falls to the bottom of the cartridge. This action continues until the steel wool is completely dissolved and the cartridge is exhausted.

The last statement is important: It is the *acid in the fixer solution* that *exhausts the cartridge,* not the amount of silver. The quantity of silver recovered can vary from 0 to 200 troy ounces (more than 13 pounds). The steel wool in a cartridge is sufficient to recover the silver in 220 gallons of fixer overflow from any KODAK Industrial X-OMAT Processor if the fixer is replenished at the recommended rate. Occasionally, an exhausted cartridge will yield more recovered silver than expected, because once a nucleus of silver has collected, it attracts more silver. However, although a cartridge may continue to collect silver after it is exhausted, the rate of collection is lower and recoverable silver is lost; it does not pay to use an exhausted cartridge.

It is also important to note that this process depends on the fixer being acidic. This is not a problem when the overflow from an X-OMAT Processor is passed through the cartridge, but it may be a problem if the fixer used in manual processing is treated.

Since all chemical reactions require time for completion, the flow of fixer from an automated processor should not exceed 300 mL per minute, and the flow should not exceed 300 mL per minute when batches of fixer from manual processing are treated; otherwise, the solution does not stay in the cartridge long enough for optimum recovery.

KODAK Chemical Recovery Cartridges are capable of efficiently removing the silver from 220 gallons of normally used KODAK INDUSTREX Fixer and Replenisher, or from 220 gallons of fixer prepared with KODAK X-ray Fixer and KODAK Rapid Fixer. If, as a result of excessive replenishment, the fixer solution has had less than normal use, the cartridges may be exhausted somewhat sooner, because the solution is more acidic and the steel wool is dissolved more rapidly. If the fixer has had more than normal use, it may be somewhat less acidic, and the cartridges would be exhausted more slowly but at a sacrifice in efficiency of silver recovery.

KODAK Silver Estimating Test Books offer a convenient, rapid method for determining exhaustion of the cartridge, as well as providing a very rough guide for determining the concentration of silver in the solutions fed into the cartridge. The books contain strips of test paper that indicate by a change in color the approximate concentration of silver in the solution. The test strips are not sufficiently sensitive, particularly at higher concentrations of silver, to serve as a substitute for chemical analysis.

When records show that a cartridge is approaching the exhaustion point (after 100 or so gallons of fixer have been passed through it), the waste solution going to the drain should be checked periodically for silver content. Merely dip a piece of test paper into the waste solution, shake the strip to remove excess liquid, and lay it on a clean sheet of white paper. After about 15 seconds, match the color of the moistened test strip with that of one of the patches on the color chart packaged with the test papers (under regular tungsten illumination). If the test indicates that the solution contains as much as 1 gram of silver per litre (⅛ ounce per gallon), the cartridge is exhausted and should be replaced. The frequency of testing depends on the rate at which the fixer is replaced (high replenishment, several times a week). Experience has shown that because of processing variables the test papers are a more accurate indicator of exhaustion than records of replenishment.

The approximate concentration of silver in a used solution is also determined by dipping a test strip into the solution and matching its color with that of a patch on the color chart. It is a simple matter to calculate the approximate yield of silver expected from a given quantity of solution.

In summary, use of the test papers is recommended to make sure that the cartridges are replaced at the proper time, that is, at the point of exhaustion.

Full details on the installation and operation of this silver-recovery system are given in Kodak Pamphlet J-9, "Silver Recovery with the KODAK Chemical Recovery Cartridge, Type P," available on request. More information about silver recovery from used processing solutions is contained in Kodak Publication J-10, "Recovering Silver from Photographic Materials."

CHAPTER 14

Special Radiographic Techniques

"IN-MOTION" RADIOGRAPHY

In most industrial radiography, it is essential that there be no relative motion of radiation source, specimen and film. Movement of the film with respect to the specimen results in a blurred radiograph; motion of the radiation source with respect to the specimen and film is equivalent to the use of a larger source size, resulting in increased geometric unsharpness.

There are certain cases, however, in which motion between components of the radiographic system—source, specimen and film—has positive benefits, either economic or in producing information that could not be otherwise obtained.

Extended Specimens

An example of an extended specimen would be a longitudinal weld in a cylindrical structure. The length of the weld could be radiographed using a series of individual exposures. This would require setup time—placement and removal of films, placement of penetrameters and identification markers, and movement of tube—for each individual exposure. Economic gains can often be achieved by radiographing the entire length of the weld in a single exposure either on a strip of film or on a series of overlapping sheets.

The x-ray beam is restricted to a narrow angle by means of a diaphragm at the tube. The tube is then traversed the length of the weld (Figure 85), each segment of weld being radiographed only during the time that the beam is incident on it. If there were two or more longitudinal welds, all could be radiographed at once using a rod-anode tube giving a 360-degree radiation beam, the radiation being restricted to a relatively thin "sheet" by lead disks concentric with the axis of the tube. In such a case, it is essential that the anode traverse along the axis of the specimen so that equal densities are obtained on the radiographs of all the welds.

The exposure time—that is, the length of time the tube is operating—is long compared to the total exposure time for a series of individual exposures. The economic advantage arises from the very great savings in setup time, and hence in total time re-

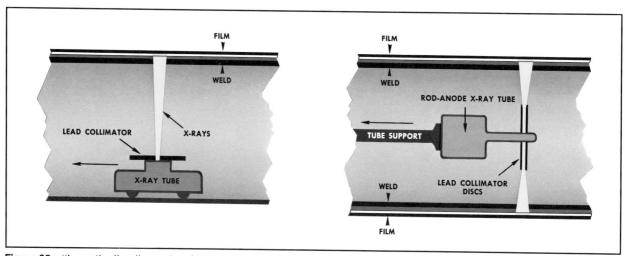

Figure 85—"In-motion" radiography of long welds. **Left:** Single weld. **Right:** Simultaneous radiography of several welds, using a rod-anode tube with disk-shaped collimators.

quired for the examination.

It is for this reason, also, that the technique is applicable largely, if not entirely, to x-rays. In gamma-ray radiography, the exposure times are usually considerably longer than the setup times. The prime advantage of the method is a saving in setup time, and it is not attractive when setup time is a small part of the total. Also, the jaws required to restrict a gamma-ray beam to a narrow angle would be very thick, especially for cobalt 60 radiation, and hence difficult to make and to position.

The technique is generally limited to thin specimens or cases where a large source-film distance can be used. The motion of the tube gives rise to a geometrical unsharpness that can be troublesome for thick specimens or short source-film distances. This unsharpness (U_m) can be calculated from the formula:

$$U_m = \frac{t\,w}{d}$$

where t is the thickness; w is the width of the radiation beam at the source side of the specimen, measured in the direction of motion of the tube; and d is the source-specimen distance.

If the permissible motion unsharpness can be estimated, the formula can be rearranged to give the maximum width (in the direction of motion) of the beam at the source side of the specimen:

$$w = \frac{U_m d}{t}$$

In the above formulae, distances are usually measured in inches and U_m is also in terms of inches. Thus, care must be exercised in comparing values so obtained with values of geometric unsharpness (U_g) calculated from the formula on page 19. Because focal spot sizes are usually specified in millimetres, values of U_g from the formula on page 19 are usually in millimetres also.

Since the exposure time T that is required for a single exposure of the specimen is usually known from experience or from an exposure chart of the material, the required rate of travel V of the tube relative to the specimen can be calculated:

$$V = \frac{w}{T}$$

where w is the width of the beam (in the direction of travel) at the source side of the specimen.

For best results with this technique, the motion of the radiation source must be smooth and uniform. Any unevenness of motion results in parallel bands of overexposure and underexposure at right angles to the direction of motion.

Rotary Radiography of Annular Specimens

Annular specimens often present economic problems if they must be radiographed in quantity. Placed flat on a film, they are wasteful of film area, of exposure and setup time, and of filing space. Great economies of time and money can be achieved by the use of a variation of the technique described above.

Film is wrapped around a circular, cylindrical, lead-covered mandrel, and covered with a light-tight covering that is also rugged enough to protect the film from abrasion. Annular specimens are then slipped on the mandrel, over the film in its protective cover. Clamping means may be required to prevent rotation of the specimens with respect to the mandrel. The "loaded" mandrel is placed behind a lead shield containing a narrow slit at least as long as the mandrel (Figure 86). During the course of the exposure, the mandrel is rotated behind the slit, each part of the specimens being radiographed in turn by the thin "sheet" of radiation passing through the slit. It is not necessary that

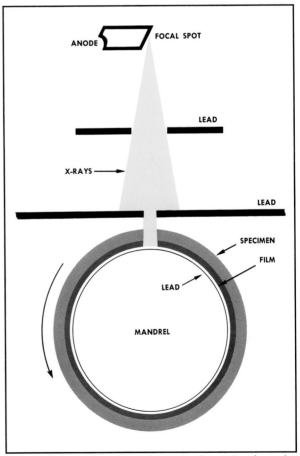

Figure 86—Plan of a setup for rotary radiography of annular specimens.

109

radiography be completed in a single turn of the mechanism. Several rotations can be used but it is important that it be an integral number of turns; otherwise there will be a band of higher or lower density on the finished radiograph. By the same token, the mandrel should rotate uniformly. Uneven rotation will cause a series of longitudinal density variations in the final record.

The lead shield should be comparatively thick. Any radiation transmitted by the shield will have the same effect on radiographic quality as additional scattered radiation in the same amount. A rule of thumb is that the shield should transmit, during the entire exposure, no more than a few percent of the amount of radiation received by a particular area of the film during the comparatively short time it is receiving an imaging exposure.

This form of radiography, like that described in the previous section, is also largely or entirely confined to x-radiation. With the average gamma-ray sources, exposure times would be long; the shield required would be very thick, requiring careful alignment of source and slit.

The motion unsharpness U_m to be expected can be calculated from the same formula (page 109):

$$U_m = \frac{t\,w}{d}$$

where w is here the width of the slit in the shield.

The required peripheral velocity V_p of the mandrel can be calculated from the exposure time T for a "still" radiograph from the expression below:

$$V_p = \frac{w}{T}$$

The same caution about units should be observed as outlined on page 109.

Scanning Methods—Orthogonal Projection

Sometimes it is desirable to use scanning methods even with specimens small enough to be radiographed in their entirety in a single exposure. This is true when critical measurements of dimensions or clearances must be made from radiographs.

When radiation passes through a specimen at an angle, as shown in Figures 87 (left) and 12, spatial relationships are distorted and measurements of clearances can be significantly falsified. If, on the

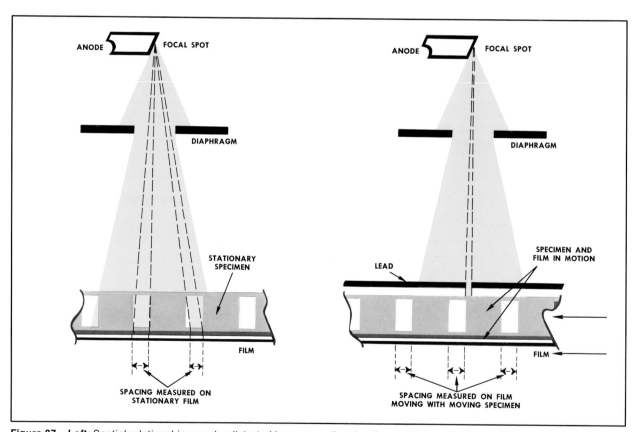

Figure 87—Left: Spatial relationships can be distorted in a conventional radiograph (see also Figure 12, page 17). **Right:** Spatial relationships are preserved by moving specimen and film through a thin "sheet" of vertically directed radiation. (For the sake of illustrative clarity, the source-film distance is unrealistically short in relation to the specimen thickness.)

other hand, the whole radiograph is made with a thin "sheet" of perpendicularly directed radiation, dimensions can be measured on the radiograph with considerable accuracy (Figure 87, right).

This is often most easily accomplished by moving the film and specimen beneath a narrow slit, with the x-ray tube rigidly mounted above the slit. The carriage carrying the film and specimen must move smoothly to avoid striations in the radiograph. Occasionally it is advantageous to move slit and tube over the stationary specimen and film. In general, however, this requires moving a greater weight and bulk of equipment, with the resulting increase in power required and in difficulties encountered in achieving smooth, even motion. Weight can be minimized by locating the slit near the focal spot, rather than close to the specimen. The disadvantage to this, however, is that the slit must be exceedingly narrow, very carefully machined, and precisely aligned with the central ray.

It should be recognized that this technique corrects for the errors of distance measurement shown in Figure 87 in only one direction—that parallel to the direction of movement of the specimen (or of tube and slit). Fortunately, most specimens to which this technique is applied are quite small, or quite long in relation to their width. If, however, the specimen is so large that measurements must be made in directions parallel to the slit (that is, at right angles to the motion) and remote from the center line, it may be necessary to scan the specimen twice in directions at right angles to one another. In a few cases, the specimen can provide its own slit. An example would be the measurement of end-to-end spacing of cylindrical uranium fuel pellets in a thin metal container. In this application, little radiation could reach the film unless it passed between the pellets almost exactly parallel to the faces of the adjacent cylinders. In such a case the slit shown in Figure 87 (right) can be comparatively wide.

Tomography

Tomography—in medical radiography, often termed "body-section radiography"—is a technique that provides a relatively distinct image of a selected plane in a specimen while the images of structures that lie above and below that plane are blurred. It is fairly common in medical radiography, and has a few specialized applications in industrial radiography.

The principle can be understood by reference to Figure 88, which shows one of the simpler and more frequently used methods. The tube and film holder are linked at the ends of a lever pivoted at the level of the plane it is desired to render. The film and tube move horizontally during the course of the exposure. As the focal spot moves to the right from X_1 to X_2 and the film to the left from F_1 to F_2, the image (P_1) of the point P remains stationary with respect to the film. However, the images of points above (A) and below (B) the point P move with respect to the film (from A_1 to A_2 and B_1 to B_2) and hence are blurred on the radiograph as shown in Figure 89.

The excursion of tube and film determines the thickness of the layer that is sharply imaged. Small motions render structures within a relatively thick section of the specimen; large motions render a thin section in sharp "focus."

Figure 88 shows a simple linear motion of tube and film, for the sake of illustrative clarity. In actual practice, the motions are often more complicated—circular, spiral, or hypocycloidal.

RADIOGRAPHY OF RADIOACTIVE MATERIALS

Examination of radioactive materials is complicated by the fact that the film is exposed to the emission of the specimen as well as to the imaging radiation. The exposure from the specimen itself is usually uniform over the whole film area, and thus

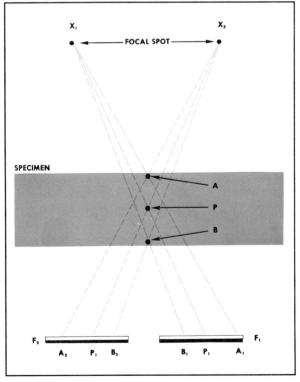

Figure 88—Basic principle of tomography. The source-specimen distance has been unrealistically shortened and the specimen-film distance increased, in order to show the principles of the system more clearly.

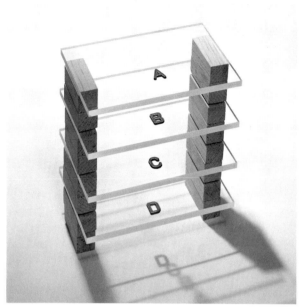

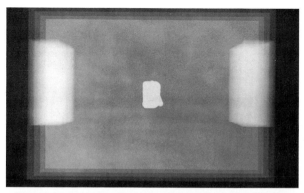

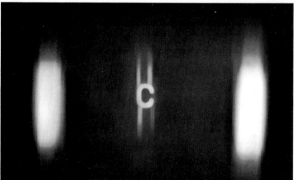

Figure 89—Demonstration of the effect obtained with tomography. The test object (above) is a series of plastic "shelves," each holding a lead letter. The first radiograph (top, right) was made with conventional radiographic equipment. The second radiograph (bottom, right), made with tomographic equipment, shows level C clearly, while the other levels are blurred.

has effects similar to, and in addition to, scattered radiation from the specimen. Best radiographic results are obtained when the proportion of this "fogging" radiation to the total radiation (fogging plus imaging) affecting the film is made as small as possible.

In general, the only radiation from the specimen that needs to be considered is the gamma radiation; any beta or alpha radiation can easily be absorbed in any material between the specimen and the film. This material may be the coating, cladding or container of the radioactive material, separate filters introduced between specimen and film, the front of the exposure holder, or any front screen used.

A number of techniques are available for minimizing the deleterious effects of the gamma radiation on the final radiograph. Which, or what combination, of these techniques is used depends on the requirements of the inspection and particularly on the activity of the specimen. As activity increases, more measures, and more complicated ones, must be employed. Because examinations of this type are highly specialized, no general directions can be given. Rather, a list of techniques, ranging from the simpler to the more expensive and complex, will be given, not all of which are necessarily applicable to every radiographic prob-

lem. It should be emphasized again that the consideration should be directed not to reducing the *absolute amount* of fogging radiation, but rather to reducing the *proportion* it bears to the imaging radiation.

Speed of Operation The most obvious way to minimize the effect of radiation from the specimen itself is to work quickly. The film is exposed to fogging radiation during the setup, exposure, and takedown times, but to the imaging radiation only during the exposure time. It is therefore advisable that the film and specimen be together for as little time as possible before exposure commences, and that they be separated as quickly as possible after the exposure has been completed.

Filtration In some cases, the gamma radiation from the specimen is of longer wavelength (that is, of lower energy or softer) than the radiation used for radiography. Under these circumstances, the fogging radiation can be considerably reduced by the use of a metallic filter between specimen and film.

Often this filtration is supplied by the lead screens used, or can be provided by the use of a thicker-than-normal front screen. Under other circumstances, separate filters may be more convenient and give better results. Specific rules are dif-

112

ficult to formulate, and the optimum filtration is best determined experimentally under the actual radiographic conditions. More commonly, the gamma emission from the specimen is at least as penetrating as the radiographic radiation, and filtration will do no good and will probably result in a decrease of radiographic quality in the image.

Radiation Intensity The proportion of fogging radiation to imaging radiation decreases as the *intensity* of the imaging radiation increases, that is, as the duration of the radiographic exposure decreases. This implies that the x-ray tube should be operated at its maximum milliamperage and that source-film distance should be as small as the requirements of image sharpness and field-coverage permit. Increasing kilovoltage may result in an improvement, but this must be checked by practical tests under operating conditions. Increasing kilovoltage decreases duration of the exposure, tending to improve the radiographic quality. On the other hand, increasing kilovoltage decreases subject contrast, which is already adversely affected by the fogging radiation from the specimen itself. Which effect predominates depends on the other factors of the examination, and thus the optimum kilovoltage must be determined empirically. If gamma rays are used for radiography, the output and specific activity of the source should be as high as possible. With x-radiography of radioactive specimens, source-film distance should be the shortest that other considerations permit.

Specimen-Film Distance Sometimes, geometric considerations of image formation (page 118) allow an increase in subject-film distance. Where such is the case, radiographic contrast is improved because when a radioactive specimen is in position A (Figure 90), some of its radiation misses the film—radiation which would affect the film in the image area if the specimen were in position B. Since the source-film distance is the same in both cases shown in Figure 90, the intensity of the image-forming radiation is unaffected. This technique results in some enlargement of the image, but often this is acceptable.

Film Speed The use of the slowest film possible reduces the relative photographic effect of the fogging radiation from the specimen, particularly when setup and takedown times are appreciable. As these times decrease, the value to be gained from using a slower film likewise decreases.

In examinations in which the specimen emits hard gamma rays and the radiography is performed with softer radiation, the use of slower types of industrial x-ray films has a second advantage. In general, the slower x-ray films have a larger ratio of sensitivities to soft and to hard radiations. This means that, relative to their sensitivities to the softer image-forming radiation, the slower films have lower sensitivities to the hard gamma radiation from the specimen.

Slit Exposures When radioactive materials are radiographed by conventional techniques, every point on the film receives fogging gamma radiation from every part of the specimen, throughout the whole exposure. The net effect of the fogging radiation can be reduced by putting a deep lead slit between the specimen and film, and traversing the slit and the x-ray tube together along the length of the specimen during the course of exposure (Figure 91). The net result of such an arrangement is that each area of the film receives only the fogging radiation directed vertically downward from that part of the specimen directly above it, rather than from the whole specimen. Note that the principle involved is very similar to that of the Potter-Bucky diaphragm (page 43). This method results in a very

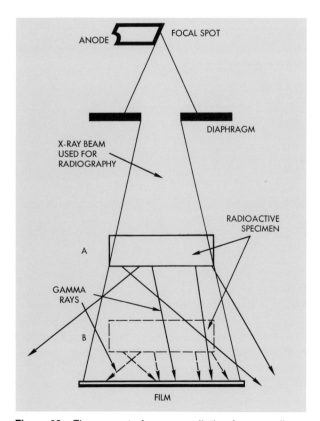

Figure 90—The amount of gamma radiation from a radioactive specimen which reaches the film depends upon the subject-film distance. Much of the radiation which "misses" the film when the specimen is in position A strikes the film in the image area when the specimen is at B.

113

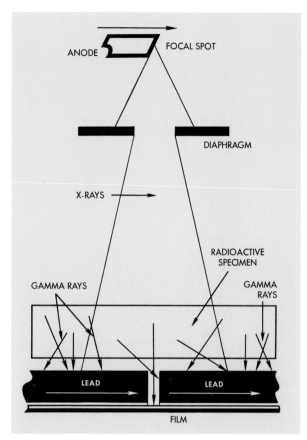

Figure 91—Schematic diagram of slit method for radiography of radioactive specimens.

great reduction in the fogging radiation and a concomitant great improvement in radiographic quality. The equipment required is often very heavy, however, since several inches of lead may be required to absorb the radiation from a highly active specimen emitting hard gamma rays. A second disadvantage is that the exposure time is considerably extended, as is the case for any scanning system of radiography (see page 108). In many cases, however, this disadvantage is of minor importance compared with the value of the examination.

Neutron Radiography Neutron radiography is probably one of the most effective methods for radiography of radioactive specimens. Its applications, however, are limited by the relative scarcity of suitable neutron sources and by the small cross-sectional areas of the available neutron beams. The techniques of neutron radiography are discussed in greater length on pages 118 and 119.

DEPTH LOCALIZATION OF DEFECTS

Two general methods are available for determining the location in depth of a flaw within a speci-

men—stereoradiography and the parallax method. The chief value of stereoradiography lies in giving a vivid three-dimensional view of the subject, although with the aid of auxiliary procedures it can be used for the actual measurement of depth. The more convenient scheme for depth measurement is the parallax method in which from two exposures made with different positions of the x-ray tube, the depth of a flaw is computed from the shift of the shadow of the flaw. Although stereoradiography and the parallax method are essentially alike in principle, they are performed differently. It is therefore necessary to discuss them separately.

Stereoradiography

Objects viewed with a normal pair of eyes appear in their true perspective and in their correct spatial relation to each other, largely because of man's natural stereoscopic vision; each eye receives a slightly different view, and the two images are combined by the mental processes involved in seeing to give the impression of three dimensions.

Because a single radiographic image does not possess perspective, it cannot give the impression of depth or indicate clearly the relative positions of various parts of the object along the direction of vision. Stereoradiography, designed to overcome this deficiency of a single radiograph, requires two radiographs made from two positions of the x-ray tube, separated by the normal interpupillary distance. They are viewed in a stereoscope, a device that, by an arrangement of prisms or mirrors, permits each eye to see but a single one of the pair of stereoradiographs. As in ordinary vision, the brain fuses the two images into one in which the various parts stand out in striking relief in true perspective and in their correct spatial relation.

The radiograph exposed in the right-shift position of the x-ray tube is viewed by the right eye, and the one exposed in the left-shift position is viewed by the left eye. In fact, the conditions of viewing the radiographs should be exactly analogous to the conditions under which they were exposed; the two eyes take the place of the two positions of the focal spot of the x-ray tube, and the radiographs, as viewed in the prisms or mirrors, occupy the same position with respect to the eyes as did the films with respect to the tube during the exposures. The eyes see the x-ray representation of the part just as the x-ray tube "saw" the actual part (Figure 92).

The stereoscopic impression is much more distinct if the specimen has a well-defined structure extending throughout its volume. If such a structure does not exist as, for example, in a flat plate of homogeneous material, it is necessary to provide such a structure upon one or more surfaces of the specimen. A widely spaced array of wires mounted on both front and rear surfaces of the specimen will generally suffice, or a similar pattern can be applied in the form of cross lines of lead paint. In stereoscopic radiographs, these added structures not only help to secure satisfactory register of the two films but also serve as a reference marking for the location of any details shown stereoscopically within the specimen.

The stereoscopic method is not often utilized in industrial radiography, but occasionally it can be of some value in localizing defects or in visualizing the spatial arrangement of hidden structures.

Double-Exposure (Parallax) Method

Figure 93 gives the details of this method. Lead markers (M_1) and (M_2) are fastened to the front and back, respectively, of the specimen. Two exposures are made, the tube being moved a known distance (a) from F_1 to F_2 between them. The position of the images of the marker (M_2) will change very little,

perhaps imperceptibly, as a result of this tube shift, but the shadows of the flaw and marker (M_1) will change position by a larger amount.

If the flaw is sufficiently prominent, both exposures may be made on the same film. (One exposure "fogs" the other, thus interfering somewhat with the visibility of detail.) The distance of the flaw above the film plane is given by the equation

$$d = \frac{bt}{a + b}$$

where d = distance of the flaw above the film plane,
a = tube shift,
b = change in position of flaw image,
t = focus-film distance.

If the flaw is not sufficiently prominent to be observed easily when both exposures are made on the same film, two separate radiographs are necessary. The shadows of the marker (M_2) are superimposed and the shift of the image of the flaw is measured. The equation given in the preceding paragraph is then applied to determine the distance of the flaw from the film.

Often it is sufficient to know to which of the two

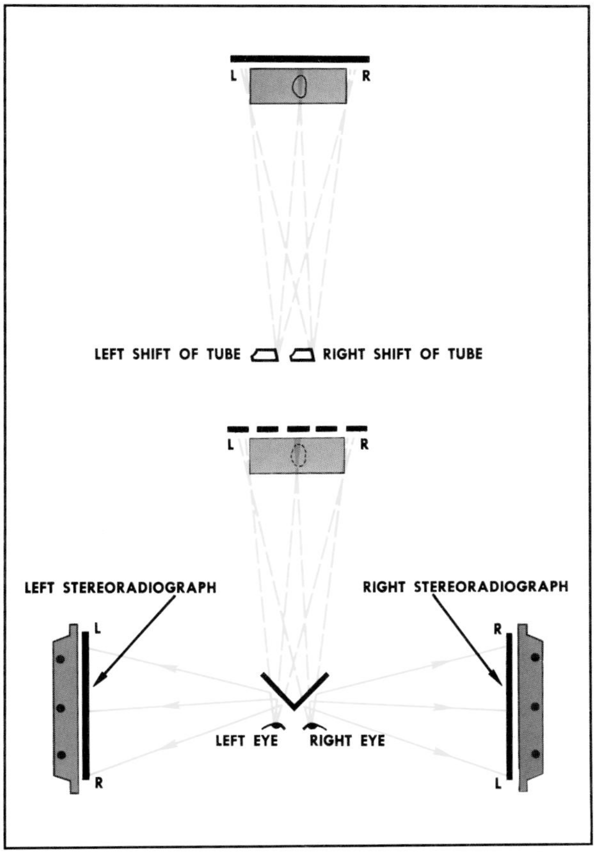

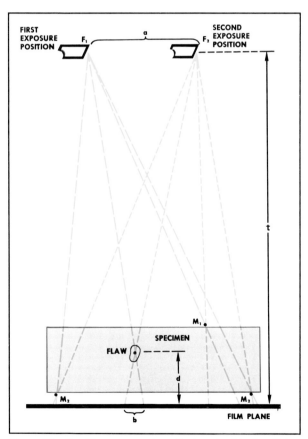

Figure 92—Above: Schematic diagram showing method of making stereoscopic radiographs. **Below:** Diagram of a stereoscopic viewer (Wheatstone type).

Figure 93—Double-exposure (parallax) method for localizing defects.

surfaces of the part the flaw is nearer. In such cases, the shifts of the images of the flaw and the marker (M_1) are measured. If the shift of the image of the flaw is less than one-half that of the marker (M_1), the flaw is nearer the film plane; if greater, it is nearer the plane of the marker (M_1).

The above methods of calculation assume that the image of the bottom marker (M_2) remains essentially stationary with respect to the film. This may not always be true—for example, if the cassette or film holder is not in contact with the bottom surface of the specimen or in a situation where large tube shifts are used.

In such cases, a graphical solution is convenient. It can be shown that if the markers are placed fairly close to the flaw, the image shifts are proportional to the distances from the film plane. Thus, a straight line graph can be drawn of image shift against distance from the bottom (film side) of the specimen.

To illustrate, let us assume that the image of the bottom marker (M_2) on a specimen 2 inches thick moved $^1/_{16}$ inch and that of the top marker (M_1) moved $^1/_4$ inch. A graph similar to that in Figure 94 is drawn, with image shift on the horizontal axis and specimen thickness on the vertical. Points M_1 and M_2 are plotted at the appropriate image shift and thickness values, and a straight line is drawn between them. (Note that for convenience in plotting, 0 specimen thickness is considered to be on the film side of the specimen.) If the image of a discontinuity has shifted, for example, $^3/_{16}$ inch, its distance from the film side of the specimen—$1^5/_{16}$ inch—can be read directly from the graph as shown in the dashed line of Figure 94. It should be emphasized that Figure 94 is not a general curve, but is for illustration of the method only. A graph of the same type must be drawn for each particular set of circumstances.

Note that in this method it is not necessary to know either source-film distance or amount of tube shift. This often makes the method convenient even in those cases where the image of the bottom marker (M_2) shifts only imperceptibly. The point (M_2) of Figure 94 then coincides with the origin of the graph.

THICKNESS MEASUREMENT

Occasionally, it is necessary to measure the thickness of a material in a location where it is difficult or impossible to use gauges, calipers, or the like. In these cases, radiography can sometimes be applied. One example of this would be the mea-

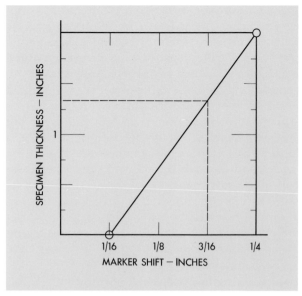

Figure 94—Example of the graphical method for localizing a discontinuity by the parallax (double-exposure) technique.

surement of the wall thickness of a hollow aircraft-propeller blade.

The technique is simple, involving the simultaneous radiography of the item in question and a stepped wedge of the same material. The stepped wedge should be wide or, if narrow, should be masked with lead in order to avoid difficulties and errors due to the undercut of scattered radiation into the area of the stepped wedge. The stepped wedge should be placed as close as possible to the area to be measured to avoid errors from differences in radiation intensity across the field. Ideally, the stepped wedge and the specimen should be radiographed on the same film. If this is not possible, as for instance when a film is inserted into a hollow propeller blade, the stepped wedge should be radiographed on film from the same box. Care should be taken that the screens used to radiograph specimen and wedge are the same. If separate films are used for wedge and specimen, they should be processed together. These precautions avoid errors caused by differences in x-ray exposure and errors in processing.

A curve is drawn relating thickness of the stepped wedge to the density obtained. Thicknesses in the part under test are then determined by reference to this curve. Plotting a calibration curve for each separate radiograph may seem laborious, but it prevents invalidation of the measurement by an unnoticed or unavoidable variation in radiographic technique.

The kilovoltage used should be as low as considerations of exposure time permit, for the sake of increased subject contrast, resulting in a higher accuracy of the measurements. By the same token, radiographs should be exposed to the highest density the available densitometer can read reliably. Because the contrast of industrial x-ray films increases with density (see pages 136 to 139), the accuracy of the thickness measurement likewise increases with density.

Special precautions should be observed if the specimen is narrow, or if measurements are to be made near the edges of a large specimen. In these cases, the specimen should be surrounded with a lead mask (see page 39), to prevent the unattenuated primary radiation from striking the film near the boundaries of the specimen. Otherwise, large errors can arise in the thickness determinations.

Determination of the composition of parts is a special application of this technique. For example, it might be possible for parts identical in appearance to be made of different alloys. In general, though not invariably, different alloys have different radiation absorptions. Thus, if a part of known composition is radiographed along with a part, or group of parts, being checked, a difference in density between corresponding areas of the control and specimens indicates a difference in composition. Note that identical densities merely make it highly probable, though not certain, that the compositions are the same. This test, of course, provides no information on heat treatment, crystal size, or the like.

It is much more difficult to estimate the size of a void or inclusion in the direction of the radiation. The density of the radiographic indication of a void depends not only on its dimensions along the direction of the beam, but also on its location within the thickness of the material and on its shape. Thus, it is necessary to prepare, by a separate experiment, calibration curves of void size versus density for each of several locations within the depth of the specimen, and perhaps for each of several shapes of void. In addition, the location of the void must be determined by one of the methods described on pages 114 to 116, in order to know which of the above-mentioned calibration curves should be used.

Such estimates of the dimensions of inclusions can be further complicated if the chemical composition and density, and hence the radiation absorption, of the inclusion is unknown.

This procedure for measuring the dimensions of voids has been applied successfully but, because of the extreme care and large amount of preliminary work required, has been limited to specimens of

high value and to circumstances where all parts of the radiographic process could be kept under the most rigorous control.

HIGH-SPEED RADIOGRAPHY

Exposure times of one-millionth of a second, or even less, can be achieved by the use of specially designed high-voltage generating equipment and x-ray tubes. Such exposure times are sufficiently short to "stop" the motion of projectiles, high-speed machinery, and the like (see Figure 95).

This apparatus differs from the usual industrial x-ray equipment in design of both the high-voltage generator and the x-ray tube. The generator contains large high-voltage condensers which are suddenly discharged through the x-ray tube, giving a high-voltage pulse of very short duration. The x-ray tube has a cold cathode rather than the conventional heated filament. Emission from the cold cathode is initiated by means of a third electrode placed near it. When electron emission has started, the discharge immediately transfers to the target which is of conventional design. The tube current may reach a value as high as 2000 amperes (two

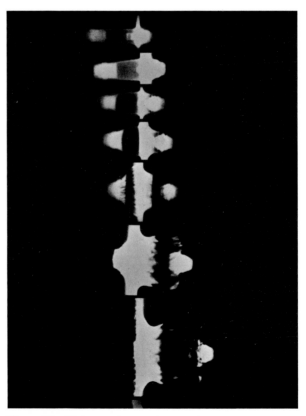

Figure 95—High-speed x-ray pictures of the functioning of a 20mm HEI shell.

117

million milliamperes), but because of the extremely short times of exposure, the load on the focal spot is not excessive.

GEOMETRIC ENLARGEMENT

In most radiography, it is desirable to have the specimen and the film as close together as possible to minimize geometric unsharpness (see page 19). An exception to this rule occurs when the source of radiation is extremely minute, that is, a small fraction of a millimetre, as for instance in a betatron. In such a case, the film may be placed at a distance from the specimen, rather than in contact with it. (See Figure 96.) Such an arrangement results in an enlarged radiograph without introducing objectionable geometric unsharpness. Enlargements of as much as three diameters by this technique have been found to be useful in the detection of structures otherwise invisible radiographically. Geo-

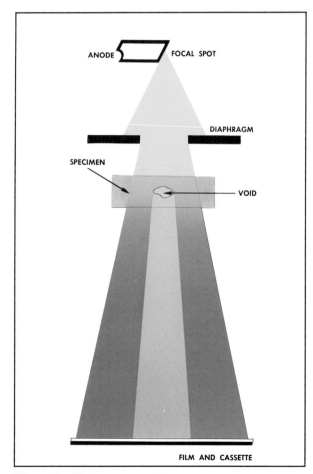

Figure 96—With a very small focal spot, an enlarged image can be obtained. The degree of enlargement depends upon the ratio of the source-film and source-specimen distances.

metric enlargements of several tens of times are feasible in microradiography (page 125).

NEUTRON RADIOGRAPHY

Neutron radiography makes use of the differential absorption of neutrons—uncharged nuclear particles having about the mass of the nucleus of a hydrogen atom (proton)—rather than of electromagnetic radiation. Neutrons, in particular those traveling at very low velocities (thermal neutrons), are absorbed in matter according to laws that are very different from those that govern the absorption of x-rays and gamma rays. The absorption of x-rays and gamma rays increases as the atomic number of the absorber increases but this is not the case with thermal neutrons. Elements having adjacent atomic numbers can have widely different absorptions, and some low atomic number elements attenuate a beam of thermal neutrons more strongly than some high atomic number elements. For example, hydrogen has a much higher neutron attenuation than does lead. Thus, the height of the water in a lead standpipe can be determined by neutron radiography—which is impossible with x-ray or gamma-ray radiography.

Neutrons, being uncharged particles, interact with matter to only a very slight degree. As a result, photographic materials are very insensitive to the direct action of neutrons and, if neutrons are to be used for imaging, some method must be used to convert their energy into a form more readily detectable photographically.

Two general classes of techniques are used in neutron radiography, both involving what are known as "converter foils." In the first—the "direct exposure" technique (see Figure 97, top)—the film is exposed between two layers of foil of a material that becomes radioactive when exposed to neutrons. The exposure to the film is caused by the beta or gamma radiation emitted by the converter foils. Foils of gadolinium, rhodium, indium, and cadmium have been used. Screens of gadolinium (0.00025-inch front, 0.002-inch back) or a front screen of 0.010-inch rhodium and a back screen of 0.002-inch gadolinium give satisfactory image quality and are among the fastest combinations.

In the second—the "transfer exposure" method (see Figure 97, bottom)—the converter foil *alone* is exposed to the neutron radiation transmitted by the specimen, and is thus rendered radioactive. After the exposure is terminated, the converter foil is placed in intimate contact with the film and the pattern of radioactivity on the converter foil produces an imaging exposure on the film.

Converter foils of gold, indium, and dysprosium have been used in this technique. A dysprosium foil about 0.010 inch thick appears to give the highest speed, and a gold foil about 0.003 inch thick appears to give the best image quality. For efficiency, both of time and of use of the radioactivity in the foil, foil and film are kept in intimate contact for three to four half-lives of the foil activity.

Conventional industrial x-ray films are ideal for neutron radiography because the photographic material is exposed to the gamma or beta radiation from the converter foil. The relative speed values for moderately hard x-radiation are satisfactory approximations for neutron radiography.

The most satisfactory neutron source for neutron radiography is a nuclear reactor which has provisions for bringing a beam of thermal neutrons to the face of the reactor shield. Neutron intensities are high enough to permit realistically short exposure times. However, unless special arrangements are made, the beams are limited to a few inches in diameter. Thus, a large object must be radiographed in sections. A few electronic and isotopic sources of thermal neutrons are available, but these suffer from the disadvantages of low neutron intensity, large effective source size, or both.

Neutron radiography is suitable for a number of tasks impossible for conventional radiography—for example, the examination of great thicknesses of high atomic number material. Several inches of lead can be radiographed with neutrons using exposure times of a few minutes. The transfer exposure method is well adapted to the radiography of highly radioactive materials such as irradiated nuclear fuel elements. Since the converter foil is unaffected by the intense beta and gamma radiation given off by the fuel element itself, the image is formed only by the differential attenuation of the neutrons incident on the element. The disadvantages of neutron radiography are primarily associated with the comparative rarity and great cost of nuclear reactors and, to a somewhat lesser degree, the small diameter of available neutron beams.

AUTORADIOGRAPHY

An autoradiograph is a photographic record of the radioactive material within an object, produced by putting the object in contact with a photographic material. In general, autoradiography is a laboratory process applied to microtome sections of biological tissues which contain radioisotopes,

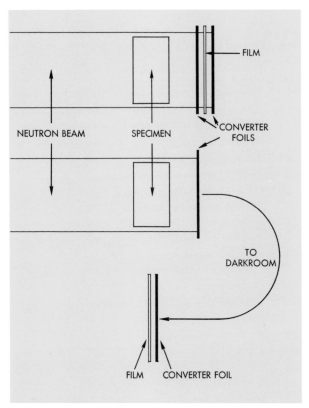

Figure 97—Two methods of neutron radiography. **Top:** Direct method. **Bottom:** Transfer method.

metallographic samples, and the like. Highly specialized techniques and specialized photographic materials—including liquid emulsions—are usually required.

However, certain autoradiographic techniques resemble those used in industrial radiography. Almost entirely limited to the nuclear energy field, they include the determination of the fuel distribution and of cladding uniformity of unirradiated fuel elements, and the measurement of fission-product concentration in irradiated fuel elements.

The first two applications involve comparatively low levels of radioactivity, and usually require the fastest types of x-ray film. The fuel element is placed in intimate contact with a sheet of film. The exposure time must be determined by experiment, but may be several hours. In the case of a nonuniformly loaded fuel element with uniform cladding, the densities recorded can be correlated with concentration of radioactive material in the element. In the case of a uniformly loaded plate, density can be correlated to the thickness of cladding. In either case, calibration exposures to one or more fuel elements of known properties are necessary, and ideally a calibration exposure should be processed with each batch of autoradiographic exposures.

119

If the nuclear fuel is unclad, a large part of the exposure to the film is caused by beta radiation. The thinnest material that gives adequate mechanical and light protection should be used between the specimen and the film. It is essential, however, that this material be exceedingly uniform in thickness. Variations in thickness will cause differences in electron transmission, and the result can easily be an "electron radiograph" of the protective material rather than a record of concentration of radioactive material in the specimen.

If the fuel element is clad, the exposure to the film is almost entirely the result of gamma radiation. Conventional exposure holders or cassettes can be used, and lead foil screens often provide substantial savings in time.

The autoradiographic determination of fission-product concentration in an irradiated fuel element usually involves exceedingly high degrees of radioactivity. The slower types of industrial x-ray film are most suitable. As in the radiography of radioactive materials, this technique places a great premium on bringing film and specimen together quickly at the start of the exposure and separating them quickly at its termination (see page 111). As in other radiographic measurement techniques, calibration exposures or rigid control of all exposure and processing variables is needed. Some means is also usually needed to protect the film or the exposure holder from radioactive contamination during the exposure to the irradiated fuel element. Thin plastic sheeting, which can be discarded after one use, has been found suitable. This, of course, is in addition to the elaborate personnel-protection measures—beyond the scope of this publication—that are also necessary.

DUPLICATING RADIOGRAPHS

Simultaneous Radiography

If it is known beforehand that a duplicate radiograph will be required, the easiest and most economical way of obtaining it is to expose two x-ray films of the same type simultaneously in the original exposure. Thus two essentially identical radiographs are produced at little or no extra cost in exposure time.

If lead-foil screen techniques are used, it is usually advisable to put both films between a single pair of screens. If each film is placed between a pair of screens, the absorption of the back and front screens, which separate the two films, results in a lower density on the back film. When two films are used between a single pair of screens, an increase in exposure is normally required because each film receives the intensification from only a single screen instead of from two. The exposure increase required must in most cases be determined by tests because it depends upon both the kilovoltage and the specimen.

If envelope-packed film with integral lead oxide screens is used, two envelopes can be superimposed since the absorption of the lead oxide layers within the package is relatively low.

With direct exposure techniques, both films should be put into the same exposure holder, either without interleaving paper or with both sheets in the same interleaving folder. With direct exposure techniques it may sometimes be found, especially at the higher kilovoltages, that exposure time must be slightly *decreased* to achieve the same density as obtained on a single film. This is because each film acts as an intensifying screen for the other. The effect is small, but may be puzzling if unexpected.

Two or more films in factory-sealed envelopes may be exposed simultaneously. No adjustment of exposure is required because the intensifying radiation from the films is largely absorbed in the material of the envelopes.

In direct exposure techniques, if it should be necessary to use separate cardboard or plastic exposure holders for each film, care should be taken to remove any lead backing from the *front* exposure holder. The presence of the lead will cause a marked difference in density between films.

In general, observance of these rules results in two identical radiographs of equal density.

Copying Radiographs

When copies of existing radiographs are needed, they can be produced by contact printing on special duplicating film. This is a direct-positive film which produces a duplicate-tone facsimile.

The characteristic curve of a typical duplicating film is shown in Figure 98. The negative slope indicates that increasing the exposure to light *decreases* developed density. Thus, if a sheet of this film is exposed to light through a radiograph and then processed, the dark parts of the radiograph (which transmit little light to the duplicating film) are reproduced as high densities and the low density areas of the original are reproduced as low densities.

Further, the gradients (see page 137) of the duplicating films are −1.0 over their useful density range. This means that, within this range, density *differences* in the original radiograph are faithfully reproduced in the copy.

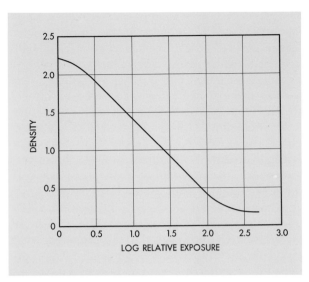

Figure 98—Characteristic curve of a typical film used for duplicating radiographs.

Exposure The exposure technique for duplicating film is simple and may be varied to suit the equipment available. The radiograph to be copied is put in a glass-fronted printing frame (available from dealers in photographic supplies). A sheet of duplicating film is placed on top of the radiograph, emulsion side in contact with the radiograph to be copied. The back of the printing frame is closed and an exposure to light is made through the glass. Any convenient light source can be used, and thus exposure conditions are difficult to specify. However, a conventional fluorescent x-ray illuminator is a usable light source. At a distance of about 2 feet between illuminator and printing frame, exposures range from several seconds to a minute. These exposures are long enough to time conveniently with simple means yet short enough to be efficient.

When a satisfactory exposure time has been found by trial for a particular radiograph, exposures for others can be estimated much more accurately. The densities in the areas of interest of the two radiographs are read, and subtracted one from the other. The antilogarithm (see page 48) of this density difference is the *factor* by which the original exposure must be multiplied (if the second radiograph is darker than the first) or divided (if the second radiograph is the lighter).

Care should be taken to keep the glass front of the printing frame free from dirt, because anything opaque adhering to the glass appears as a dark mark on the processed duplicate.

Reproduction of Densities As shown by the characteristic curve of Figure 98, the maximum density obtainable (that is, with *no* exposure on the dupli-cating film) may be below the maximum density in the radiograph to be copied. Thus, if the densities in the areas of interest in an industrial radiograph were 3.0 and 3.5, the duplicating film, the curve of which is shown in Figure 98, could not reproduce these densities. It would, however, reproduce the density *differences* exactly, and thus reproduce exactly the *radiographic contrasts* of the original. If the densities in the original were 3.0 and 3.5, these could be reproduced as 0.7 and 1.2, respectively, or 1.0 and 1.5, or 1.3 and 1.8, depending upon the exposure given the duplicating film.

In many cases, a reproduction of the radiographic contrasts alone is quite sufficient. (Note that if a radiograph with densities of 1.0 and 1.5 were displayed on an illuminator, and a similar radiograph but with densities of 3.0 and 3.5 were displayed on an illuminator 100 times as bright, an observer would be unable to distinguish between them.)

Sometimes, as when sets of reference radiographs are being prepared, it is required to reproduce both the densities and the density differences of the original. This can be done by mounting a uniform density filter behind the copy. In the example cited above, the densities of the original were 3.0 and 3.5, and the densities of the copy were 1.0 and 1.5. If a uniform density of 2.0 is added, the total densities will be raised to 3.0 and 3.5, just as in the original.

The least expensive and most convenient neutral density filter is a processed sheet of fine-grained photographic film which has been uniformly exposed to light. This film should be coated on clear base, so that the color of the copy is not changed. Further, the film chosen should be very slow so that exposure is easy to control. Ideally, it should also be sensitive only to the blue portion of the visible spectrum and may be handled under the safelighting conditions used for x-ray films.

In assembling copies and neutral density filters, the duplicating film and the neutral density filter should be positioned so that their emulsion sides are toward the viewer.

FLUOROSCOPY

Fluoroscopy differs from radiography in that the x-ray image is observed visually on a fluorescent screen rather than recorded on a film. A diagrammatic sketch of an industrial fluoroscopic unit is shown in Figure 99.

Fluoroscopy has the advantages of high speed and low cost.

121

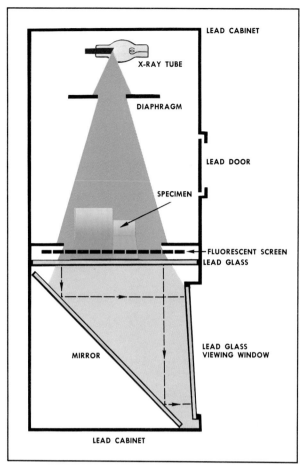

Figure 99—Schematic diagram of an industrial fluoroscope. Commercial models may differ from the illustration. For more rapid examinations, industrial fluoroscopes may be provided with material conveyors.

However, fluoroscopy has three limitations: (1) Examination of thick, dense, or high-atomic-number specimens is impractical, because the x-ray intensities passing through them are too low to give a sufficiently bright image on the fluorescent screen. (2) The sensitivity of the fluoroscopic process is not as great as that of radiography. This is caused in part by the lower contrast and coarser grain of the fluoroscopic screen as compared to the film record, and in part by the relatively short source-screen distances that must be used to obtain high screen brightnesses. This latter factor also increases the distortion of the fluoroscopic image. (3) The lack of a permanent record of the examination may be a further disadvantage.

The main application for fluoroscopy is in the rapid examination of light, easily penetrated articles, the unit value of which does not warrant the expense of radiography, or of items for which a highly sensitive test is unnecessary. Fluoroscopy

has been used, for example, in the inspection of packaged foods for foreign objects and of molded plastic parts for the correct placement of metallic inserts. In some cases, it is advantageous to sort parts fluoroscopically before they are radiographed to save the expense of radiographing specimens which contain gross flaws.

An extension of fluoroscopy involves the use of image intensifiers (Figure 100). In these, the x-rays, after traversing the specimen, strike a fluorescent screen (the "input phosphor"). The fluorescence of the screen causes the photoelectric surface with which it is coated to emit electrons in proportion to the intensity of the fluorescence. These electrons are accelerated and focused by electrostatic lenses onto a second fluoroscopic screen (the "output phosphor") much smaller than the first. The second phosphor has a brightness several hundreds of times that of the first, partly because of its smaller size and partly because of the additional energy imparted to the electrons by the accelerating voltage in the image intensifier. The second phosphor can be viewed directly by means of a suitable optical system. Alternatively, the image on the output phosphor may be picked up by a television camera and displayed on a television monitor at any convenient location. The use of a television link permits the brightness and the contrast of the final image to be adjusted independently of any radiographic variables. The use of image intensifiers

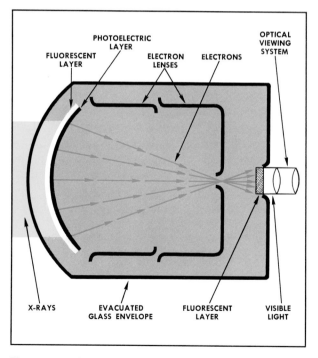

Figure 100—Schematic diagram of a fluoroscopic image intensifier.

where they are applicable avoids many of the visual difficulties attributable to low screen brightness.

In a third method which may be classified as fluoroscopy, the functions of image intensifier and television camera are combined. The sensing element is an x-ray-sensitive television pickup tube, the output from which is fed through a video amplifier to one or more television monitors at remote locations. The sensitive areas of the pickup tubes are often rather small, making the system applicable only to the examination of small items such as electronic components or spot welds, or of narrow subjects such as longitudinal welds in thin materials. On the other hand, the small size of the x-ray-sensitive area, coupled with the fairly large size of the television display, results in a direct magnification of the image that may be as great as 30 diameters. With this equipment as well, brightness and contrast of the final image can be adjusted electronically.

PHOTOFLUOROGRAPHY

In photofluorography (Figure 101) the image on the fluorescent screen is photographed with a camera on small or even miniature film rather than viewed directly. In medicine, the economy of this procedure has made it useful in the examination of large groups of people for disease of the chest. In rendition of detail, photofluorography is superior to fluoroscopy because the film can be viewed with ample illumination; the photon exposure can be integrated to a high exposure level; and the photographic process enhances the contrast of the fluorescent image on the screen.

Compared to full-sized radiography, the factors that diminish detail in the miniature photoradiograph are the graininess and diffusion in the fluorescent screen, the limitations of the lens in definition, and the relatively greater influence of film graininess in the small image. Since requirements of medical and industrial radiography are not the same, photofluorographic units for each application may be expected to differ from one another in many details. In the industrial field, photofluorography has been used for the inspection of parts for which the sensitivity requirements are not severe, and where the value of the part is too low to permit the expense of conventional radiography.

A modification of photofluorography is cinefluorography—the production of x-ray motion pictures. In the simplest form of cinefluorography, the still camera shown in Figure 101 is replaced by a motion-picture camera. This form of the technique

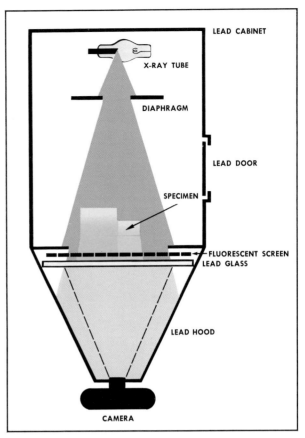

Figure 101—Schematic diagram of an industrial photofluorographic unit. Details of commercial units may differ from this illustration.

is limited to relatively thin specimens, to low frame rates, or to both, because the available fluorescent-screen brightness is restricted by the permissible x-ray tube loads. Cinefluorography can be extended to thicker specimens or higher frame rates by the use of an image intensifier (page 122) in place of the simple fluoroscopic screen. Cinefluorography is essentially a research tool, and as a consequence the details of the apparatus depend very strongly on the requirements of the particular investigation.

MICRORADIOGRAPHY

Sometimes—usually with thin specimens of low x-ray absorption—the detail required for study is too fine to be seen with the naked eye, and examination of the image through a low-powered microscope or enlargement of the radiograph by ordinary optical projection, is required to visualize the available detail. This method, *microradiography*, generally employs soft x-rays generated in the range of 5 to 50 kV. The photographic emulsion is usually single coated and finer grained than the emulsion of ordinary x-ray films.

Commercial applications of microradiography have included such unrelated projects as studying cemented joints in corrugated cardboard, and distinguishing between natural and cultured pearls. Biological materials, such as tissue sections, insects, and seeds, have been examined. In horticulture, the distribution of inorganic spray materials on foliage has been investigated by means of microradiography. Of particular interest to the metallurgist is the demonstration of minute discontinuities and the segregation of the constituents of an alloy in a thin section. (See Figure 102.)

Both the continuous, or "white," x-radiation and the characteristic K spectrum from a suitable target find use in microradiography. The continuous spectrum can be used for detection of minute discontinuities or of segregation in alloys in which the components differ fairly widely in atomic number—for example, the examination of aluminum-copper alloys (Figure 103), or in the determination of the dispersion of lead in a leaded brass. For such applications, tungsten-target, beryllium-window tubes operating up to 50 kV are useful. The continuous spectrum from x-ray diffraction tubes can also be used. Microradiography has also been accomplished using the white radiation from ordinary radiographic tubes operated at a low voltage, although the x-ray intensity obtained at low voltages is severely limited by the thickness of the tube window.

When segregation of components that do not differ greatly in atomic number must be detected,

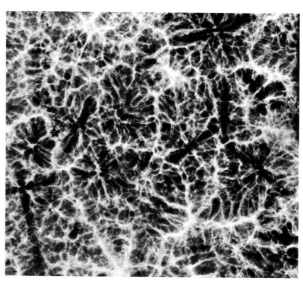
Figure 103—Microradiograph of a cast aluminum alloy containing about 8 percent copper, made on a slow photographic plate of low graininess. Light areas are copper-rich. Enlargement is 35 diameters.

the use of the characteristic K x-ray spectrum from a suitable element gives the best results. K characteristic radiation can be obtained by two methods. An x-ray tube with a target of the suitable material has the advantage of a relatively high intensity of K radiation, but has the disadvantage of requiring several x-ray tubes of different target materials or a tube with demountable targets. In addition, a number of elements that emit K radiation that might be useful cannot be made into x-ray tube targets. Use of K fluorescence radiation, obtained by irradiating a secondary target of a suitable material with the intense continuous spectrum of a tungsten-target tube avoids the disadvantage of the first method and gives a nearly pure K spectrum. However, the intensities available by this method are relatively low and long exposure times are needed.

Extremely good contact between specimen and film is necessary if the maximum enlargement of which the film is capable is to be achieved. Good contact can be obtained with a simple mechanical jig which presses the specimen against the film or plate, but best results are probably obtained with a vacuum exposure holder. This consists of a plate with a milled recess into which the photographic material and specimen are placed and over which is put a flexible x-ray transparent cover. Evacuation of the recess with a vacuum pump or water aspirator causes the atmospheric pressure on the cover to press the specimen into intimate contact with the film or plate. (See Figure 104.)

Figure 102—Microradiograph of leaded brass made on a very slow photographic plate of low graininess. Enlargement is about 150 diameters.

Any material placed between tube and specimen must be thin and of low atomic number (for example, a thin sheet of cellulose derivative) to minimize x-ray absorption, and must have no marked structure. Certain sheet plastic materials contain chlorine. These should not be used as covers for microradiographic film holders because their absorption for soft x-rays is likely to be high.

The source-film distance usually ranges from 3 to 12 inches. The choice of tube voltage or K-emitting element must be based on the character of the specimen and may be quite critical if the finest radiographic results are to be obtained from a particular specimen. The type of film or plate must be selected to meet the degree of magnification to be used in examining the radiograph.

It is of little value to specify exact exposure techniques in microradiography, but the following example may serve as a useful guide: Specimen, plate of aluminum-base alloy ¼ mm thick; exposure, 20 kV, 25 mA, 12-inch source-film distance; time,

1½ minutes using the slowest available industrial x-ray film.

Because of the relatively high absorption by air of the very soft x-radiation used in microradiography, the x-ray intensity decreases with distance from the focal spot more rapidly than calculations based on the inverse square law would indicate (page 46).

In conventional microradiography, the specimen and film must be in close contact because the subsequent optical enlargement of the radiograph demands that geometric unsharpness (see page 19) be minimized. By the use of special x-ray tubes having focal spots about 1 micron (0.00004 inch) in diameter, the source-specimen distance may be made very small and the specimen-film distance some tens of times greater. This results in a geometric enlargement in the original microradiograph which can often be viewed directly without any intervening optical or photographic steps.

Enlargement of Microradiographs

If a microradiograph must be enlarged more than 50 diameters, conventional photomicrographic techniques and equipment must be used. However, it is seldom necessary to enlarge industrial microradiographs more than 50 diameters. Such enlargements may be prepared by a technique known as photomacrography. The term *photomacrography* refers to enlargements of about 50 diameters or less, made with a single lens, as opposed to photomicrography which implies greater enlargements, using a compound microscope. Photomacrographic equipment is relatively inexpensive, and is often readily available in plant photographic departments.

Lenses and Cameras The simplest method for obtaining moderate enlargements of microradiographs, and one that frequently gives satisfactory results, is to use a conventional photographic enlarger. A microfilm enlarger embodying a "point source" of light and condenser illumination is preferred, but any photographic enlarger of good quality may be used. The microradiograph is placed in the negative carrier of the enlarger and a sheet of photographic paper or film is placed on the easel. As in all photographic enlarging, it is important that the enlarger be rigid; if the negative carrier of the enlarger vibrates with respect to the easel during the exposure, the enlarged image will be blurred.

For enlargements up to 50 diameters, best results are obtained if the specialized equipment—cameras, lenses, and specimen holders—available for photomacrography is employed. Several optical

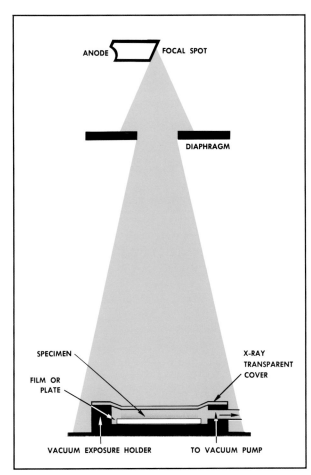

Figure 104—Arrangement for microradiography using a vacuum exposure holder. The specimen thickness has been exaggerated for the sake of illustrative clarity.

manufacturers make lenses specially for this purpose. View cameras or those specially designed for enlarged-image photography are most convenient, although an enlarger with a film adapter back can also be used.

In addition, short focal-length camera or enlarger lenses can be employed if they are used "backward"—that is, with the front of the lens toward the film in the camera and the back of the lens toward the microradiograph being reproduced. The disadvantage of using these lenses is that the aperture settings are inside the camera and can be changed only by removing the lens from the camera or by reaching down the bellows. This disadvantage may be outweighed by the easy availability of the lenses.

When a lens of 16 mm focal length is used, a 50-diameter enlargement requires a lens-to-film distance of about 32 inches. Longer focal-length lenses require greater bellows extension on the camera for the same enlargement. Lower magnifications can be obtained either by reducing the lens-to-film distance, or by using a lens of a different focal length.

A properly focused condenser lens is preferred for illuminating the microradiograph. Ideally, the condenser lens should be chosen to suit the camera lens being used. The illumination is properly adjusted when an image of the lamp filament is in focus on the diaphragm leaves of the camera lens, and slightly larger in size than the camera-lens aperture that will be used when making the enlargement. A setup for photomacrography is illustrated in Figure 105. A heat-absorbing filter *must be used* to prevent buckling of the microradiograph.

If a specimen stage (specimen holder) and condenser lens are not available, the microradiograph may be placed on opal glass or ground glass and illuminated from behind. Opal glass is preferable to ground glass, but both cause some loss in image quality, particularly at the greater magnifications. Because diffuse illumination of the microradiograph wastes a great deal of light compared to condenser illumination, exposure times can be as much as 20 times longer than with condenser illumination. In addition, the camera image, being dim, is much more difficult to focus accurately.

Photographic Films and Papers The enlarged reproduction of the microradiograph may be either a positive (tone scale reversed from that of the original) or a negative (tone scale the same as that of the original). The choice depends on the use to be made of the enlargement and, in some cases, on the density scale of the original microradiograph.

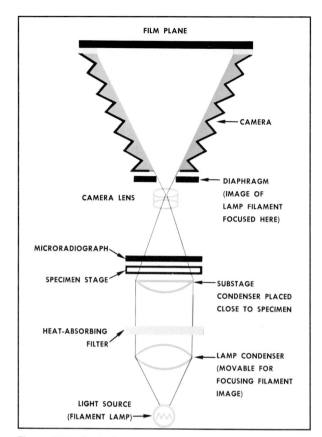

Figure 105—Optical arrangement for photomacrography.

Positive enlargements are simpler to make than are negative enlargements because they involve the minimum number of photographic steps. For low magnifications, they are also quicker, provided that only one or two enlargements of each microradiograph are needed. However, many microradiographs contain too great a range of densities to be reproduced satisfactorily in a single photographic step. The useful scale of photographic papers is much less than that of microradiographic films or plates. Hence, if a microradiograph containing a wide range of densities is projected directly onto a photographic paper, details may be lost in either the high or low density areas or in both. Thus, direct printing is only useful for microradiographs of relatively low contrast. A density difference of about 1.2 to 1.4 between the highest and lowest densities of interest in the microradiograph is about the maximum density scale that can be accommodated by the lowest contrast photographic enlarging paper (Grade 1). Microradiographs with lower density scale can be printed on papers of higher contrast as shown in Table VII.

The densities of very small areas on microradiographs may be difficult or even impossible to measure with the result that trial exposures on several grades of paper may be necessary. The criterion, whether the grade of paper is chosen from Table VII or by trial and error, is that the densities in the area of interest in the microradiograph just "fill up" the exposure scale of the paper.

TABLE VII—PAPER GRADES FOR REPRODUCTION OF MICRORADIOGRAPHS

Enlarging Paper Grade Number	Density Scale of Microradiograph or Intermediate
1	1.2 to 1.4
2	1.0 to 1.2
3	0.8 to 1.0
4	0.6 to 0.8
5	below 0.6

Often, the density scale of interest in the microradiograph is greater than 1.4, or a duplicate-tone reproduction is required, or many copies of the enlarged microradiograph are needed. It is then necessary to make an intermediate copy on film at the desired magnification, and to contact-print this intermediate on a suitable grade of paper.

Because of the wide range of densities in most microradiographs a low-contrast intermediate is usually necessary. A fine-grained sheet film of the type used for portraiture, developed to a gamma of about 0.4, is generally satisfactory.

By contact printing this intermediate, copies of the microradiograph can be conveniently produced in any desired quantity. The grade of paper to use with any intermediate can be chosen from Table VII or by trials with several grades of paper. Grade 0 is available as a contact-printing paper, and will satisfactorily reproduce intermediates where the density range is 1.4 to 1.6. Loss of significant detail can result if some areas on the paper print are overexposed to a uniform black and/or others so underexposed that they present a uniform white. Fitting the paper grade to the area of interest in the intermediate insures that significant details appear with the maximum contrast, but without loss of "highlight" or "shadow" detail.

ELECTRON RADIOGRAPHY

In electron radiography, electrons emitted by lead foil irradiated by x-rays pass through a thin specimen of low atomic number. They are differentially absorbed in their passage and record the structure of the specimen on a film (Figure 106).

Specimens that can be examined by electron radiography are limited by the range of the electrons to thin, light materials. Papers, wood shavings, leaves, fabrics, and thin sheets of rubber and plastic have been examined by this method.

A conventional front lead foil screen, 0.005 inch thick, is a suitable source of electrons. The x-rays used should be generated at the highest kilovoltage possible, up to at least 250 kV and a filter equivalent to several millimetres of copper should be placed in the tube port. The very hard x-radiation is needed because the electron emission of lead foil screens

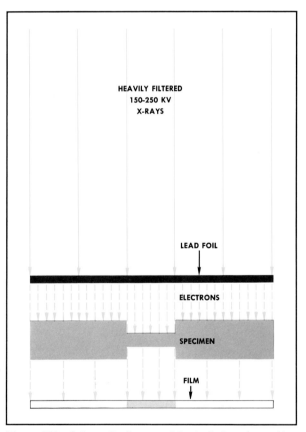

Figure 106—Schematic diagram of the technique for making electron radiographs. More electrons can reach the film through the thin portions of the specimen than through the thick portions. (For illustrative clarity, the electron paths have been shown as straight and parallel; actually, the electrons are emitted diffusely.)

increases with increasing hardness of radiation, up to several hundred kilovolts. In other words, as the penetration of the radiation increases, the photographic effect of the electrons from the lead foil screens becomes greater relative to the photographic effect of the direct x-rays. In electron radiography, the useful image is formed only by electron action, and the direct x-rays act only to produce a uniform overall exposure. It is therefore desirable to minimize the relative intensity of the contrast-reducing direct radiation by achieving the highest electron emission—that is, the highest intensification factor—possible.

The slowest industrial x-ray films are most suitable for electron radiography. When the film used is double-coated, it is desirable to protect the back emulsion from the action of the developer or to remove it after processing (see page 104) because this emulsion contains no image, only the uniform density resulting from direct x-ray exposure.

Maximum enlargements feasible in electron radiography are much lower than those possible in microradiography with the same film. This is because the electrons are strongly diffused in the specimen, and also because film graininess increases markedly in going from the voltages used in microradiography to the high energy x-ray and electron exposures used in electron radiography (see pages 67 and 145).

Good contact between lead foil, specimen, and film is essential. The electron emission from lead foil is diffuse, and the electrons are further diffused as they pass through the specimen. Therefore, any space between foil, specimen, and film results in great deterioration in image sharpness (see Figure 27, page 31). A vacuum cassette (page 124) should be used whenever possible or, lacking this, screen, specimen, and film should be clamped in intimate contact. In particular, the pressure supplied by a conventional spring-backed cassette may be insufficient to produce adequate contact if the specimen—for instance, a leaf or a wrinkled piece of stiff paper—is not itself flat.

A method related to electron radiography is electron-emission radiography—sometimes referred to in Europe as "reflection microradiography." This method is based on the fact that electron emission from a substance exposed to x-rays depends upon the atomic number of material, among other factors. When a photographic or radiographic film is placed in intimate contact with a specimen and the whole irradiated *from the film side* with hard x-rays such as are used in electron radiography, differences in electron emission resulting

from differences in atomic number and variations in concentration of components are recorded on the developed film, as is shown in Figure 107.

Hard x-rays are required in this technique for the same reason as in electron radiography: It is necessary to maximize the ratio between the electron exposure, which produces the image, and the uniform overall exposure produced by the direct x-rays.

Specimens to be examined by electron-emission radiography need to be smooth and plane on one side only and can be quite massive—significant factors when the specimen cannot be cut.

An electron-emission radiograph resembles a photomicrograph in that it shows details only of the surface of a specimen whereas microradiography shows the distribution of a component throughout the thickness of the specimen. For this reason, and because electron emission is less sensitive to differences in atomic number than is soft x-ray absorption, an electron-emission radiograph

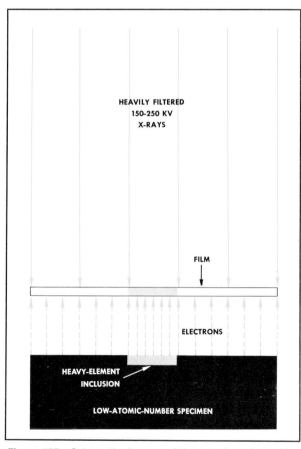

Figure 107—Schematic diagram of the technique for making electron-emission radiographs. More electrons are emitted from the areas of the specimen which contain materials of high atomic number. (For illustrative clarity, the electron paths have been shown as straight and parallel; actually, the electrons are emitted diffusely.)

of a specimen contains markedly less information than a microradiograph of a thin section of the same specimen.

Image definition is also poorer than in a microradiograph because of the diffuse emission of electrons from the surface and because of the increased graininess of the film associated with the high-energy x-ray and electron exposures. Nevertheless, electron-emission radiography is a valuable technique, for instance, in the examination of printed matter to distinguish nondestructively between inks containing metallic pigments and those containing aniline dyes.

The photographic materials used for microradiography are also useful for electron-emission radiography. However, the maximum enlargements possible are considerably less than those attained in microradiography with the same film because of the diffuse nature of the electron emission and the greater graininess. As in the case of electron radiography, the back emulsion of a double-coated film should be protected from the action of the developer or it should be removed after processing (see page 104).

It is necessary to insure good contact between specimen and film because of the diffuse emission of the electrons that produce the image. In the case of small specimens, a vacuum cassette, into which both specimen and film can be put, is very useful.

The three techniques—soft x-ray radiography (microradiography), electron radiography, and electron-emission radiography—can be compared using a postage stamp* (Figure 108A) as a specimen. The design of the stamp is green and the "Deutsches Reich" overprint is black. Figure 108B is a soft x-ray radiograph, which could be greatly enlarged if necessary, showing details of both design and paper. The image of the design is a negative indicating the absorption of the x-radiation by the ink. The electron radiograph (Figure 108C) was made with the design of the stamp away from the recording film, and contains details of the paper structure only. The "wavy line" watermark shows very clearly. The electron-emission radiograph (Figure 108D) shows the details of the design alone, indicating that the green ink has a high electron emission and hence that it contains a metallic pigment, rather than aniline dye. No trace of the black overprint is visible because the carbon-based ink has negligible electron emission.

---

*It is unlawful to make radiographs of stamps of the United States of America or other countries without specific authorization from the Chief of the U.S. Secret Service, Washington, D.C. This permission has been received.

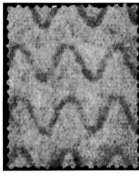

Figure 108—Bavarian stamp of 1920 (Scott No. O52). **A:** Photograph. The design is green and the "Deutsches Reich" overprint is black. **B:** Soft x-ray radiograph. Details of both design and paper are visible. Design is "negative," indicating absorption of the x-rays by the ink. **C:** Electron radiograph. Only the details of the paper are shown. **D:** Electron-emission radiograph. The design is "positive," indicating a relatively high electron emission from some heavy element in the ink. The overprint cannot be seen.

X-RAY DIFFRACTION

This application of x-rays has nothing in common with the shadow picture method ordinarily referred to as radiography. It is of considerable value, however, in the industrial laboratory for by its use compounds can be identified, crystalline structures determined, and the effect on metals of cold work and annealing studied.

A crystal is made up of a definite geometric arrangement of atoms, and the regular planes thus established are capable of diffracting x-rays. In this process of diffraction, the x-rays are deflected in different directions and form a pattern on film, the character of this pattern being determined by the arrangements of atoms and tiny crystals in the specimen. If a change is brought about in the crystalline state of the material, by heat treatment, for example, the change will be registered in the pattern. X-ray diffraction is useful in pure research as well as in manufacturing process controls.

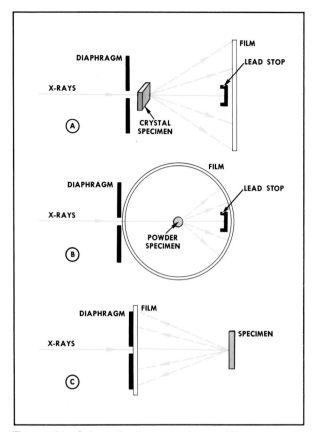

Figure 109—Schematic diagrams of x-ray diffraction methods. **A:** The Laue method uses heterogeneous radiation and a stationary crystal. In the Bragg method, the crystal is rotated or oscillated and monochromatic radiation is used. **B:** Debye-Scherrer-Hull (powder) method. **C:** Back-reflection method.

In the Laue method, a narrow beam of x-rays (Figure 109A) from the continuous spectrum passes through a single crystal of the material under investigation. The primary beam should be stopped at the film with a small piece of lead to prevent scattered x-rays from fogging the film. Some of the x-rays are diffracted, but only in particular directions, and the places where these rays impinge show as more or less dense spots on the developed film (Figure 110).

The position of the spots is determined by Bragg's Law, namely, $n\lambda = 2d \sin \theta$ where n is the order of the spectrum, λ the wavelength of the x-rays, d the distance between the atomic planes involved, and θ the angle between the diffracted x-ray and the atomic plane. Because a crystal is a regular arrangement of atoms in space, there are many planes at different angles, just as there are many different rows of trees with different spacings in an orchard, depending on the angle at which the orchard is viewed. If the crystal of Figure 109A is properly oriented and suitably rotated or

oscillated (Bragg's method), and the x-rays used are monoenergetic, the information thus obtained is particularly useful for the determination of the internal structures of crystals (see Figure 111).

Besides certain technical limitations of the Laue method, the necessity for always having a single crystal available is a handicap. The Debye-Scherrer-Hull method (Figure 109B) commonly utilizes a small amount of the sample in powdered form packed in a fine glass capillary, or stuck to a fiber or flat ribbon of material transparent to the x-rays. The resulting random distribution of tiny crystals permits Bragg's Law to be satisfied for many different lattice planes within the crystal. In this system, monoenergetic x-rays are used. Determination of crystal lattice constants can be made by the Debye-Scherrer-Hull method (Figure 112).

Some indication of the size of the crystals of the powder can be derived from the character of the diffraction lines. The method also applies to the study of metallurgical samples in the form of foil, since these are, in general, polycrystalline. If the specimen is sufficiently transparent to the x-rays, the arrangement in Figure 109B may be used; but, if it is too thick, the rays diffracted backward would be used, as in Figure 109C. In this way information is derived about mechanical

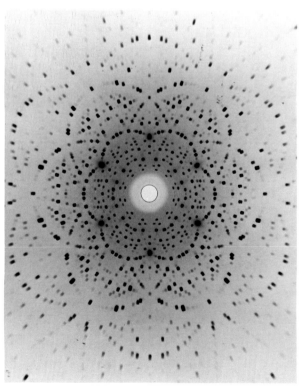

Figure 110—X-ray diffraction pattern of the mineral beryl (beryllium aluminum silicate) made by the Laue method (Figure 109A).

Figure 112—Powder diffraction patterns by the Debye-Scherrer-Hull method. **Top:** Sodium chloride (common salt) (NaCl). **Bottom:** Sodium sulphate, anhydrous (Na_2SO_4).

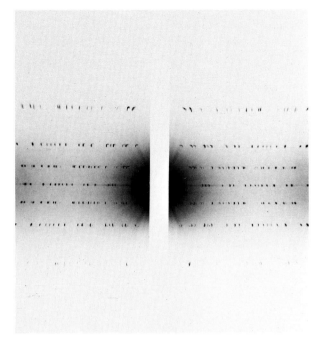

Figure 111—X-ray diffraction pattern of silver oxalate ($Ag_2C_2O_4$) made by the rotating crystal method (Figure 111A).

Often the location, rather than the intensity, of lines or spots of x-ray diffraction patterns is of importance. Under these circumstances, considerable savings in exposure time can be achieved by the chemical aftertreatment known as intensification, described on page 102.

strains and preferred orientations in the specimen, and about the relation between these indications and the nature and amount of heat treatment or of mechanical working.

Films for X-ray Diffraction

The choice of a film for a particular problem depends on the radiation quality and the relative importance of film speed, contrast, and graininess.

Since the parallax associated with an image on double-coated film may be objectionable in certain applications of x-ray diffraction, a single-coated film may be preferable. However, if a double-coated film is used, one emulsion can be protected against the action of the developer by a waterproof tape which is removed after the film is in the fixer. Alternatively the emulsion may be removed from one side of the completely processed film. Methods of applying both these techniques are described on page 104.

X-ray Diffraction References

AMOROS, JOSE L., BUERGER, MARTIN J., and AMOROS, MARISA C.—*The Laue Method*, Academic Press, New York, San Francisco, London, 1975.

AZAROFF, LEONID V., and BUERGER, MARTIN J. —*The Powder Method in X-ray Crystallography*, McGraw-Hill Book Company, Inc., New York, 1958.

JEFFERY, J. W.—*Methods in X-ray Crystallography*, Academic Press, London and New York, 1971.

KLUG, H. P., and ALEXANDER, L. E.—*X-ray Diffraction Procedures for Polycrystalline and Amorphous Materials* (2nd ed), John Wiley and Sons, New York, 1974.

LIPSON, H., and STEEPLE, H.—*Interpretation of X-ray Powder Diffraction Patterns*, St. Martin's Press, New York, 1970.

STOUT, GEORGE H., and JENSEN, LYLE H.—*X-ray Structure Determination*, The Macmillan Company, New York, 1968.

Powder Diffraction File, International Center for Diffraction Data, 1601 Park Lane, Swarthmore, Pa., 19081.

CHAPTER 15

Paper Radiography

To many radiographers and interpreters the words "radiograph" and "film" are synonymous. However, a combination of factors, among them, recurring silver shortages and rising costs of other non-destructive testing methods, have prompted increasing interest in the use of paper in industrial radiography. Developments in papers, screens and processing techniques have resulted from the realization that radiographs on paper have distinct advantages to offer the user who will consider them, not in the context of x-ray film, but as the product of another recording medium, one which is to be viewed in an entirely different way.

Advantages of Paper Radiographs

What are some of the advantages of using paper in industrial radiography? For one, *rapid access.* A damp-dry radiograph can be put in the interpreter's hands in as little as 10 seconds after exposure. Moreover, radiographic paper, plus intensifying screens, plus proper exposure equals *good image quality*. With direct exposure, the image has acceptable subject contrast combined with wide latitude. *Convenience* and *economy* also enter the picture. The paper processor is portable, requires no plumbing connections and, in addition, has a low operating cost.

Applications for Paper Radiography

Where can you use paper radiographs in your non-destructive testing program? The applications are numerous. For instance, there are many stages in production which may require radiographic inspection *before* code or specification radiographs are made. This is a prime area for the use of paper in radiography. Other applications exist in foundries where in-process x-ray procedures are used to monitor practices of gating and risering; checking core positions in wax patterns; detecting shrinkage flaws, porosity, dross, or cavities in castings; and monitoring root passes and subsequent weld passes for a variety of flaws.

Radiographs on paper also find application in checking circuit boards for proper assembly and absence of solder balls. In aircraft maintenance, paper radiographs can be used to inspect for water in honeycomb, foreign material in oil pumps, and for general survey work. Some other applications include on-site checking of pipelines, pressure vessels and weldments; inspection in industries such as food processing, wood products, tires, seeds, and titanium reprocessing; bomb detection and many types of survey radiography.

Factors Affecting Paper Radiographs

Although there are areas of similarity between radiographs on paper and radiographs on film, recognition of the differences between the two is, in some instances, the governing factor for the production of good paper radiographs. Paper requires more exposure control than film because it has less exposure latitude and a shorter density scale. On the other hand, both paper and film require control of scatter radiation in order to realize the most that is in the product.

Exposure Techniques

The fact that radiographic papers have a shorter density scale and less exposure latitude than industrial x-ray films demands more critical exposure parameters. There are, therefore, precise techniques for exposing these papers to achieve a properly exposed paper radiograph quickly. It is possible to choose a kilovoltage which will tend to yield an acceptable penetrameter sensitivity for a given metal and thickness range.

Processing Techniques

Radiographs on paper can be processed by stabilization, by automatic film processing, or manually.

STABILIZATION PROCESSING

Stabilization processing is a method of producing radiographs on paper much faster than is possible by conventional develop-stop-fix-wash processing. The primary differences between stabilization processing and ordinary radiographic processing are in the speed and mechanism of development and in the method of treating the unexposed, light-sensitive silver halide left in the emulsion after development. Exposed paper processed by stabilization makes quality, ready-to-use radiographs available in seconds. These stabilized radiographs are not permanent because the chemical reactions within the emulsion have been stopped only temporarily. They will, however, last long enough to serve a number of practical purposes.

In conventional processing, unused silver halide is dissolved by fixer and any traces of soluble silver compounds left after fixing are removed by washing. The resulting radiographs are stable for long periods. In stabilization processing, the silver halide is converted to compounds which are only temporarily stable and the radiographs have a limited keeping time. Stabilized radiographs often remain unchanged for many months if they are not exposed to strong light, high temperature, or excessive humidity. Commercial keeping quality can be achieved by fixing and washing if a longer-lasting record is desirable.

Papers designed for stabilization processing have developing agents in the paper emulsion. Development is achieved by applying an alkaline activator to the emulsion surface. The stabilizer is applied to neutralize the activator and to convert any remaining silver halide to *relatively* stable, colorless compounds. Ordinary photographic papers or x-ray films cannot be developed by this process because there are no developing agents present in either the emulsion or the activator. Stabilization papers with developing agents in the emulsion can also be hand-processed in x-ray processing chemicals.

Advantages of the Stabilization Process

In a stabilization process, a measure of radiographic stability is exchanged for some definite advantages.

Simplicity The process is adaptable to uncomplicated mechanical systems.

Space Saving Darkroom space and plumbing needs are greatly reduced. In fact, some applications of the process do not require a darkroom.

Water Saving Stabilized radiographs do not require washing.

Greater Uniformity Mechanically processed radiographs have better day-to-day uniformity in density than those processed manually.

Successful Stabilization Processing

To realize the advantages of stabilization processing, there are a number of factors which must be taken into account.

Correct Exposure This is essential because in this type of processing the developing time is constant.

Processor Maintenance Follow the manufacturer's recommendations for cleaning and maintenance of the processor.

Capacity of Solutions. Don't overwork chemical solutions Observe the manufacturer's recommendations in regard to capacity and renewal of solutions. Discard solutions—

1. When 150 square feet of radiographic paper has been processed

2. After 1 week regardless of the amount of paper processed

3. When a processed radiograph shows noticeable degradation

Dry Processing Trays Check before loading them with chemicals. Some stabilization solutions are not compatible with water.

Avoid contamination of the activator with the stabilizer This results in chemical fog on the radiographs. The smell of ammonia is an indication of contamination.

Do not wash stabilized radiographs unless they have been fixed in an ordinary fixing bath Washing without fixing makes a stabilized radiograph sensitive to light.

Because stabilized radiographs are impregnated with chemicals, *do not file* them *in contact with processed x-ray films* or other valuable material. Stabilized radiographs which are to be kept for an *extended period of time* must be *fixed and washed* (post-stabilization processing).

AUTOMATED PROCESSING

Some radiographic papers can be processed in specially modified automated film processors. However, papers designed for stabilization processing are not usually processable in film processors. Check the paper manufacturer's recommendations for specific processing information.

Radiographic paper cannot be intermixed with x-ray film for processing. Replenishment rates for paper are much lower than for film. Consequently, if film is intermixed with paper, the film will receive improper processing.

MANUAL PROCESSING

Most radiographic papers can be processed manually. Check the manufacturer's recommendations for the specific processing chemicals, times, and temperatures required.

VIEWING PAPER RADIOGRAPHS

A correctly exposed, properly processed radiograph on paper is only part of the story. To be useful in providing information, the radiograph must be viewed, and viewing radiographs on light-reflecting paper is entirely different from viewing radiographs on light-transmitting film. It is almost immediately apparent that some of the familiar methods of measurement and interpretation applicable to film are not relevant to the interpretation of paper radiographs.

Density—Transmission vs Reflection

When electromagnetic radiation—in the form of light, x-rays, or gamma rays—reacts with the sensitive emulsion of x-ray film or radiographic paper, the emulsion will show a blackening after it has been processed. The degree of blackening is defined as *density*. Up to this point, radiographic paper acts identically like film, but beyond this point, differences appear.

Density Measurement

The density on transparent-based film is known as *transmission density*, D_T, and is defined as the logarithm of the ratio of the incident light intensity, I_O (from the illuminator), on the radiograph, to the light intensity transmitted through the radiograph, I_T. The formula is:

$$D_T = \text{Log} \frac{I_O}{I_T}$$

Since this formula applies only to light-transmitting images, it cannot be applied to an opaque-based imaging material such as radiographic paper. Therefore, a slightly different means of measuring density is necessary, and this is called *reflection density*, D_R. Reflection density is defined as the logarithm of the ratio of incident light intensity, I_O, to the reflected light intensity, I_R, from the image area. This formula is:

$$D_R = \text{Log} \frac{I_O}{I_R}$$

So, although the formula appears to be quite similar to that of transmission density, in practical application, reflection density measures the light reflected from the radiograph, not that which passes through. For example, reflection densities are measured by a *reflection densitometer*, and the familiar densitometer for measuring transmitted densities cannot be used.

To carry the discussion one step further, exposure is defined as the product of the quantity of radiation—measured in roentgens or other units—and time. In this respect, the exposure to radiographic paper is measured exactly the same as it is for x-ray film, although the order of magnitude of the exposure may be different. Reflection characteristic curves can be generated for radiographic paper. The difference between these curves and those for film is that transmission density values are used for x-ray film, while reflection densities are used for radiographic paper. Characteristic curves (which are also known as H & D curves) for paper have a shorter range of densities and usually a shorter log exposure scale.

Comparable Densities—Paper and Film

Interpreters could easily be led astray at this point by becoming involved in the purely objective relationship between reflection density and transmission density. They may theorize, for example, that under a given set of viewing conditions, a reflection density of 0.7 appears to them to be similar to a transmission density of 2.0. In reality, the transmission density to reflection density relationship has no bearing upon where the same radiographic information is recorded on the film's transmission density scale as related to the paper's reflection density scale. The question really is—which densities contain the same information if x-ray film and radiographic paper are used to record the same image?

Given the correct intensity of illumination, it is universally believed that the most useful information is recorded on the essentially straight line portion of the characteristic curve. In fact, it is rather generally accepted that in industrial radiography the higher the density of a film radiograph, the better the visibility of tiny discontinuities—limited mainly by the available intensity of the illuminator. Because of the essentially opaque base of paper radiographs and the shoulder effect of the H & D curve of paper, this concept cannot be applied to reflection densities.

A radiographic image on paper of the same image area of a subject will contain the same important image details as a radiograph on film. These details will be modified in density (and possibly in contrast) because the response is fundamentally different.

It must also be recognized, of course, that the *total range* of information capable of being recorded will be less on a paper radiograph because the reflection density scale is shorter than its x-ray film counterpart. For example, a reflection density of 2.0 on a paper radiograph is so black that detail is completely obscured.

In addition, because of the opaque nature of the paper base, the method of viewing reflected densities of images on paper is fundamentally different from the method for viewing transmitted densities. Although these differences exist and must be recognized, the similarities in practical usage between film and paper radiographs are even more striking. *Good practice indicates that the exposure given to a radiographic paper is adjusted until the necessary and desirable details of the image are distributed along the available density scale of the paper within the constraints of optimum reflection viewing.*

If this is done correctly, it will be noticed that the important details will tend to be found in the mid-scale of subjective brightness provided by the density scale of the paper. This is strikingly similar to that of a film radiograph in which the details of a good image tend to be centered around the middle of the density scale (usually about 2.0). The center point, or aim point, then, is a significant factor for visualization of detail for both paper and film radiographs—even though the aim point may be a different value, and the densities may be reflected or transmitted.

INTERPRETING PAPER RADIOGRAPHS

Whether produced on film or on paper, a radiograph containing useful information must be viewed by an observer for the purpose of interpretation. The viewing process is, therefore, a subjective interpretation based on the variety of densities presented in the radiograph. To perform this function, the eye must obviously be capable of receiving the information contained in the image. Judgments, likewise, cannot be made if the details cannot be seen.

Viewing conditions are obviously of utmost importance in the interpretation of radiographs. As a general rule, extraneous reflections from, and shadows over, the area of interest must be avoided, and the general room illumination should be such that it does not impose any unnecessary eyestrain on the interpreter.

When following these general guidelines in viewing film radiographs, then, the light transmitted through the radiograph should be sufficient only to see the recorded details. *If the light is too bright, it will be blinding; if too dim, the details cannot be seen.* The general room illumination should be at approximately the same level as that of the light intensity transmitted through the radiograph to avoid shadows, reflections and undue eyestrain.

The natural tendency is to view radiographs on paper, like a photograph, in normal available light. For simple cursory examination this can be done, but since normal available light might be anything from bright sunlight to a single, dim, light bulb, some guidelines are necessary. It has been found from practical experience that radiographic sensitivity can be greatly enhanced if the following guidelines for viewing are observed.

1. As noted in the general rule, all extraneous shadows or reflections in the viewing area of the radiograph that adversely affect the eyes must be avoided. In fact, *a darkened area, minimizing ambient lighting,* is desirable.

2. Since radiographs on paper must be viewed in *reflected light, several sources of reflected light* have been used successfully. One method is the use of *specular light* (light focused from a mirror-like reflector) directed at an angle of approximately 30° to the surface of the radiograph, from the viewer's side so that reflected light does not bother the eyes. Light which comes from a slide projector is specular light.

   Other sources are the familiar high-intensity reading light, like a Tensor light, or a spotlight. Another type of light which has been found to be very effective is a *circular magnifying glass* illuminated around the periphery with a *circular fluorescent bulb.* When using this form of illumination, the paper radiograph should be inclined at an oblique angle to the light to produce the same specular lighting just discussed. These devices found in drafting rooms as well as medical examining rooms are usually mounted on some sort of adjustable stand, and have the advantage of low power magnification—on the order of 3X to 5X. The magnifying glass should be such that it does not distort the image, but it does emphasize the fact that the graininess characteristics of paper radiographs are minimal.

Reflection density is different from transmission density. The difference is important in viewing and interpreting radiographs on paper, but presents no difficulty in procedure or visualization.

CHAPTER 16 — Sensitometric Characteristics of X-ray Films

THE CHARACTERISTIC CURVE

As pointed out on page 52, the relation between the exposure and the density* in the processed radiograph is commonly expressed in the form of a *characteristic curve*, which correlates density with the logarithm of relative exposure.

Contrast

The slope, or steepness, of the characteristic curve for x-ray film changes continuously along its length. It was shown qualitatively (page 53) that the density difference corresponding to a difference in specimen thickness depends on the region of the characteristic curve on which the exposures fall. The steeper the slope of the curve in this region, the greater is the density difference, and hence the greater is the visibility of detail. (This assumes, of course, that the illuminator is bright enough so that a reasonable amount of light is transmitted through the radiograph to the eye of the observer.)

The increasing ease of visibility of detail with increasing steepness of the characteristic curve is demonstrated in Figure 113. Figures 113B and C are radiographs of the test object shown in Figure 113A. The radiographs differ only in the milliampere-seconds used to make them, that is, on the portion of the characteristic curve on which the densities fall. The details are much more clearly visible on radiograph C than on B because C falls on the steeper high-density portion of the characteristic curve (where the film contrast is high), while B falls on the much flatter toe portion.

The type of example dealt with qualitatively, above, also lends itself to quantitative treatment.

The slope of a curve at any particular point may be expressed as the slope of a straight line drawn

---

*Photographic density is dimensionless, since it is the logarithm of a dimensionless ratio. There are, therefore, no "units" of density. In this respect, it is similar to a number of other physical quantities, for example, pH, specific gravity, and atomic weight.

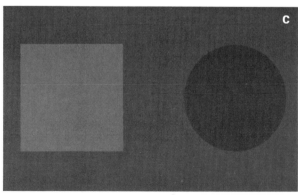

Figure 113—A: Photograph of test object. **B** and **C:** Radiographs of test object. The exposure time for C was greater than that for B, all other factors remaining constant. Note that the structure of the test object is more clearly seen in C since it was made on a steeper portion of the characteristic curve.

tangent to the curve at the point. When applied to the characteristic curve of a photographic material, the slope of such a straight line is called the *gradient* of the material at the particular density. A typical characteristic curve for a radiographic film is shown in Figure 114. Tangents have been drawn at two points, and the corresponding gradients (ratios a/b, a'/b') have been evaluated. Note that the gradient varies from less than 1.0 in the toe to much greater than 1.0 in the high-density region.

Now consider two slightly different thicknesses in a specimen. These transmit slightly different intensities of radiation to the film; in other words, there is a small difference in the logarithm of the relative exposure to the film in the two areas. Let us assume that at a certain kilovoltage the thinner section transmits 20 percent more radiation than the thicker. The difference in logarithm of relative exposure ($\Delta \log E$) is 0.08, and is independent of the milliamperage, exposure time, or source-film distance. If this specimen is now radiographed with an exposure that puts the developed densities on the toe of the characteristic curve where the gradient is 0.8, the x-ray intensity difference of 20 percent is represented by a density difference of 0.06 (see Figure 115). If the exposure is such that the densities fall on that part of the curve where the

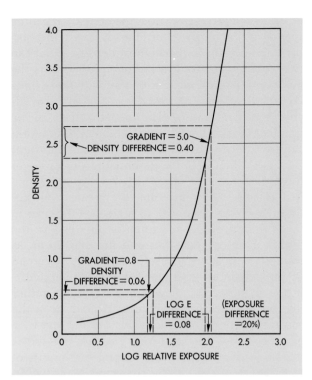

Figure 115—Characteristic curve of a typical industrial x-ray film. Density differences corresponding to a 20 percent difference in x-ray exposure have been evaluated for the two values of gradient illustrated in Figure 114.

gradient is 5.0, the 20 percent intensity difference results in a density difference of 0.40.

In general, if the gradient of the characteristic curve is greater than 1.0, the intensity ratios, or subject contrasts, of the radiation emerging from the specimen are exaggerated in the radiographic reproduction; and the higher the gradient, the greater is the degree of exaggeration. Thus, at densities for which the gradient is greater than 1.0, the film acts as a "contrast amplifier." Similarly, if the gradient is less than 1.0, subject contrasts are diminished in the radiographic reproduction.

A minimum density is often specified for radiographs. This is not because of any virtue in a particular density, but rather because of the gradient associated with that density. The minimum useful density is that density at which the minimum useful gradient is obtained. In general, gradients lower than 2.0 should be avoided whenever possible.

The ability of the film to amplify subject contrast is especially significant in radiography with very penetrating radiations which produce low subject contrast. Good radiographs depend on the enhancement of subject contrast by the film.

The direct x-ray characteristic curves of three typical x-ray films are shown in Figure 116. The gradients of these curves have been calculated, and are

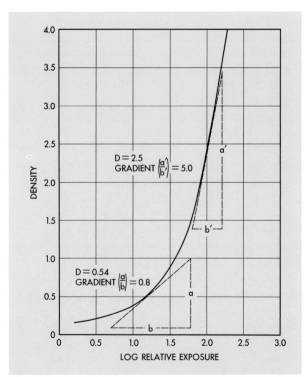

Figure 114—Characteristic curve of a typical industrial x-ray film. Gradients have been evaluated at two points on the curve.

137

plotted in Figure 117 against the density. It can be seen that the gradients of Films X and Y increase continuously up to the highest densities that can conveniently be used in radiography. This is the basis for the recommendation that with these films one should use the highest densities that the available illuminators allow to be viewed with ease. The gradient versus density curve of Film Z has a form different from the others in that the gradient increases, then becomes essentially constant over the density range of about 1.5 to 2.5, beyond which it decreases. With this film, the greatest density difference corresponding to a small difference in transmission of the specimen is obtained in the middle range of densities and the maximum, as well as the minimum, useful density is governed by the minimum gradient that can be tolerated.

It is often useful to have a single number, rather than a curve as shown in Figure 117, to indicate the contrast property of a film. This need is met by a quantity known as the *average gradient,* defined as the slope of a straight line joining two points of specified densities on the characteristic curve. In particular, the specified densities between which the straight line is drawn may be the maximum and minimum useful densities under the conditions of practical use. The average gradient indicates the average contrast properties of the film over this

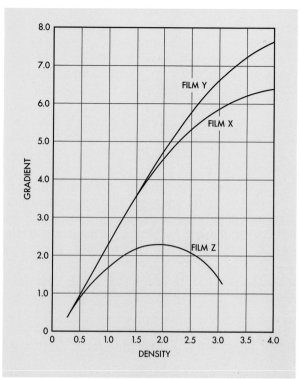

Figure 117—Gradient versus density curves of the typical industrial x-ray films, the characteristic curves of which are shown in Figure 116.

useful range; for a given film and development technique, the average gradient depends on the density range chosen. When high-intensity illuminators are available and high densities are used, the average gradient calculated for the density range 2.0 to 4.0 represents the contrast characteristics of the film fairly well. If high densities are for any reason not used, a density range of 0.5 to 2.5 is suitable for evaluation of this quantity. If intermediate densities are used, the average gradient can be calculated over another range of densities—1.0 to 3.0, for example.

Figure 118 shows the characteristic curve of a typical industrial x-ray film. The average gradients for this film over both the above density ranges are indicated. In the table below are given average gradients of the typical x-ray films, the characteristic curves of which are shown in Figure 116. Since Film Z does not reach a density of 4.0, its average gradient cannot be given for the higher density range.

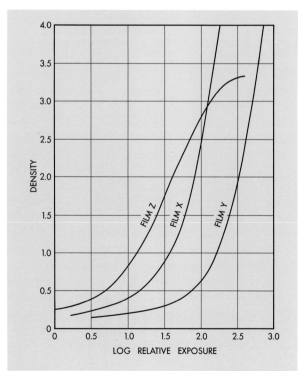

Figure 116—Characteristic curves of three typical industrial x-ray films.

	Average Gradient	
Film	Density Range 0.5—2.5	Density Range 2.0—4.0
X	2.3	5.7
Y	2.6	6.3
Z	1.7	—

138

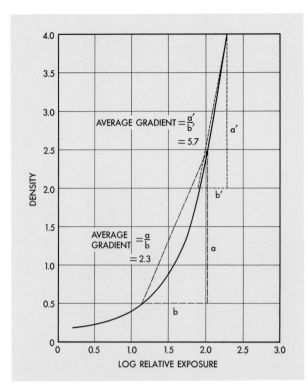

Figure 118—Characteristic curve of a typical industrial x-ray film, with the average gradient calculated over two density ranges.

Experiments have shown that the shape of the characteristic curve is, for practical purposes, largely independent of the wavelength of x-radiation or gamma radiation. Therefore, a characteristic curve made with any radiation quality may be applied to exposures made with any other, to the degree of accuracy usually required in practice, and the same is true of values of gradient or average gradient derived from the curve.

The influence of kilovoltage or gamma-ray quality on contrast in the radiograph, therefore, is the result primarily of its effect on the subject contrast, and only very slightly, if at all, of any change in the contrast characteristics of the film. Radiographic contrast can also be modified by choice of a film of different contrast, or by use of a different density range with the same film. Contrast is also affected by the degree of development, but in industrial radiography, films are developed to their maximum, or nearly their maximum, contrast. In the early stages of development, both density and contrast increase quite rapidly with time of development (see page 142). However, to use manual processing as an example, the minimum recommended development time gives most of the available density and contrast. With certain of the direct x-ray film types, somewhat higher speed and, in some cases, slightly more contrast are gained by extending the development, but in no case should the maximum time recommended by the manufacturer be exceeded.

A special case arises when, for technical or economic reasons, there is a maximum allowable exposure time, that is, exposure time cannot be increased to take advantage of a higher film gradient at higher densities. In such a case, an increase in kilovoltage increases the radiation intensity penetrating the specimen, and hence the film is exposed to a higher density. This may result in an increase in radiographic contrast. An example may be taken from the exposures used to produce the exposure chart shown in Figure 44. The following tabulation lists the densities obtained through the 1/2- to 5/8-inch sections, using an exposure of 8 mA-min.

kV	D_B 1/2" steel	D_A 5/8" steel	Radiographic Contrast $D_B - D_A$	Relative Radiographic Contrast
120	0.50	0.27	0.23	20
140	1.20	0.67	0.53	46
160	2.32	1.30	1.02	88
180	3.48	2.32	1.16	100

These data show that, when the exposure time is fixed, the density difference between the two sections increases, and hence the visibility of detail in this thickness range is also improved as the kilovoltage is raised. The improvement in visibility of detail occurs in spite of the decrease in the subject contrast caused by the increase in kilovoltage, and is the direct result of using higher densities where the gradient of the film is higher. Qualitatively, one may say that, in this particular case, the film contrast is increasing faster as a result of increased density than the subject contrast is decreasing as a result of increased kilovoltage. It should be emphasized again that this change in radiographic contrast resulting from a change in kilovoltage is not the result of a change in shape of the characteristic curve but rather the result of using a different portion of the characteristic curve—a portion where the slope is greater.

Speed

It has been shown that the film contrast depends on the *shape* of the characteristic curve. The other significant value obtained from the characteristic curve is the relative speed, which is governed by the *location* of the curve, along the log E axis, in relation to the curves of other films.

In Figure 116, the curves for the various x-ray films are spaced along the log relative exposure axis. The spacing of the curves arises from the differences in relative speed—the curves for the faster films lying toward the left of the figure, those for the slower films toward the right. From these curves, relative exposures to produce a fixed density can be read; the relative speeds are inversely proportional to these exposures. For some industrial radiographic purposes, a density of 1.5 is an appropriate level at which to compute relative speeds. However, the increasing trend toward high densities, with all radiographs viewed on high-intensity illuminators, makes a density of 2.5 more suitable for much industrial radiography. Relative speed values derived from the curves in Figure 116 for the two density levels are tabulated below, where Film X has arbitrarily been assigned a relative speed of 100 at both densities.

| Film | Density = 1.5 | | Density = 2.5 | |
	Relative Speed	Relative Exposure for D = 1.5	Relative Speed	Relative Exposure for D = 2.5
X	100	1.0	100	1.0
Y	24	4.2	26	3.9
Z	250	0.4	150	0.7

Note that the relative speeds computed at the two densities are not the same because of the differences in curve shape from one film to another. As would be expected from an inspection of Figure 116, this is most noticeable for Film Z.

Although the shape of the characteristic curve of a film is practically independent of changes in radiation quality (see page 142), the location of the curve along the log relative exposure axis, with respect to the curve of another film, does depend on radiation quality. Thus, if curves of the type of Figure 116 were prepared at a different kilovoltage, the curves would be differently spaced, that is, the films would have different speeds relative to the film that was chosen as a standard of reference.

DENSITY-EXPOSURE RELATION

The most common way of expressing the relation between film response and radiation intensity is the characteristic curve—the relation between the density and the logarithm of the exposure—which has been discussed above. If, however, density is plotted against *relative exposure* to x-rays or gamma rays—rather than against the *logarithm of the exposure*—in many cases there is a linear relation over a more or less limited density range (see Figure 119). If *net* density (that is, density above base density and fog), rather than gross density, is plot-

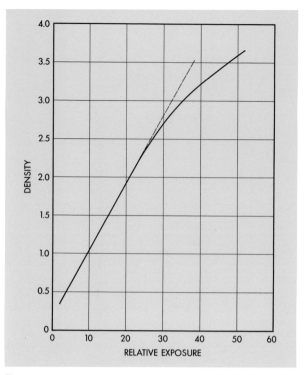

Figure 119—Density versus exposure curve for typical industrial x-ray film exposed to direct x-rays or with lead screens. The density range over which a linear relationship exists between density and exposure depends upon both film type and development.

ted against exposure, the straight line passes through the origin.

The linear relation cannot be assumed, however, but must be checked for the conditions involved in the particular application because the density range over which it is valid depends on the film used and on the processing conditions. The linear relation between density and exposure may be extremely useful, for instance, in the interpretation of diffraction patterns and the evaluation of radiation monitoring films, provided it is kept in mind that the linear range of the curve is limited.

RECIPROCITY LAW FAILURE*

The Bunsen-Roscoe reciprocity law states that the resultant of a photochemical reaction is dependent only on the *product* (I × t) of the radiation intensity (I) and the duration of the exposure (t), and is independent of the absolute values of either quantity. Applied to radiography, this means that the developed density in a film depends only on the *product* of x- or gamma-ray intensity reaching the film and the time of exposure.

---

*See also page 48.

The reciprocity law is valid for direct x-ray or gamma-ray exposures, or those made with lead foil screens, over a range of radiation intensities and exposure times much greater than those normally used in practice. It fails, however, for exposures to light and, therefore, for exposures using fluorescent intensifying screens. The magnitude of the differences that may be expected in practice are shown in the table on page 49. Figure 120 shows a conventional reciprocity curve. (The data on page 49 were derived from a curve of this type.) The vertical axis in Figure 120 has been considerably expanded to make the curvature more apparent.

The logarithms of the exposures ($I \times t$) that produce a given density are plotted against the logarithms of the individual intensities. It can be seen that for a particular intensity (I_0) the exposure ($I \times t)_0$ required to produce the given density is a minimum. It is for this intensity of light that the film is most efficient in its response. For light intensities higher (I_H) and lower (I_L) than I_0, the exposure required to produce the given density is greater than ($I \times t)_0$. Phrased differently, there is a certain intensity of light for which a particular film is most efficient in its response. In radiography with fluorescent screens, failure of the reciprocity law sometimes gives results that *appear* to be a failure of the inverse square law (see page 49).

In most industrial radiography, the brightness of fluorescent intensifying screens are very low, and the exposures used lie on the left-hand branch of the curve shown in Figure 120. An exception is high-speed flash radiography (page 117) in which the exposures lie to the right of the minimum.

EFFECT OF DEVELOPMENT TIME ON SPEED AND CONTRAST

Although the shape of the characteristic curve is relatively insensitive to changes in x- or gamma-ray quality, it is affected by changes in degree of development. Degree of development, in turn, depends on the type of developer, its temperature, its degree of activity, and the time of development. Within certain limits, increased degree of development increases the speed and contrast of an x-ray film. If, however, development is carried too far, the speed of the film as based on a certain net density* ceases to increase and may even decrease. The fog increases and contrast may decrease.

Automated Processing

Automated processors and their associated chemicals are designed to give the optimum degree of development. All the variables, listed in the above paragraph, that affect the degree of development are controlled and kept constant by the processor. The responsibilities of the operator are to keep the machine clean and to make sure that temperatures, replenishment rates, and the like are maintained at the proper levels.

Manual Processing

In manual processing, however, all the processing variables are under the control of the operator. Although *all* are important in the production of the final radiograph, those associated with development are particularly so. The following discussion is therefore devoted to the effects of development variables, in the context of manual processing.

Figure 121 shows the characteristic curves of a typical industrial x-ray film developed for a series of times in developer at 68° F. It can be seen that as development time increases, the characteristic curve grows progressively steeper (contrast increasing) and moves progressively to the left (speed increasing). Note that the curves for 2 and 10 minutes represent development techniques that would never be encountered in practice. They are included here merely as extreme examples.

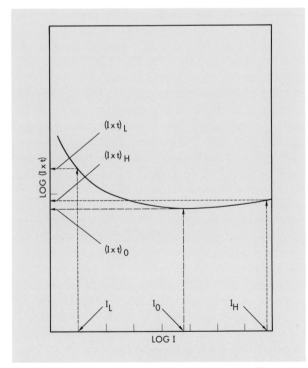

Figure 120—Reciprocity curve for light exposures. The corresponding curve for direct x-rays or lead screen exposures would be a straight line parallel to the log I axis.

—————
*Net density—density above base density and fog.

141

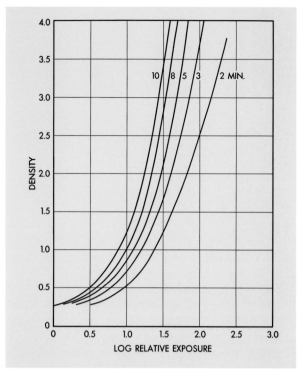

Figure 121—Characteristic curves of a typical industrial x-ray film, developed for 2, 3, 5, 8, and 10 minutes.

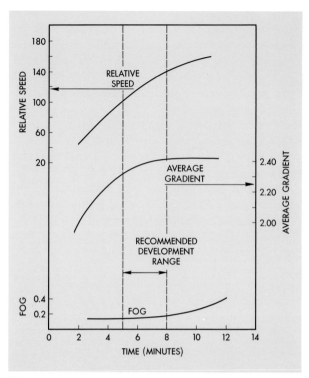

Figure 122—Time-speed and time-average gradient curves derived from the characteristic curves of Figure 121. The time-fog curve for the same development conditions is shown.

From the curves shown in Figure 121, values of relative speed and average gradient can be derived as explained on pages 138 and 140. These values, and fog, are plotted against development time in Figure 122. Curves of these quantities plotted against developer temperature, with development time held constant, show a similar form. At the maximum recommended development time of 8 minutes, the film used in this example has reached its maximum contrast, and further development would cause a decrease in film contrast, rather than an increase.

Curves of the type shown in Figure 122 depend upon the particular film considered. Some films, for instance, increase very little in average gradient when the development time is increased from 5 to 8 minutes, the main effect being an increase in film speed. A practical point, however, should be noted here. Although the average gradient of a film *(film contrast)* may be unaffected by a change of development time, increased development time may result in an increase in *radiographic contrast,* that is, density differences in the radiograph. This is illustrated by Figure 123 in which are plotted the characteristic curves of a typical industrial x-ray film for 5 and 8 minutes' development. These curves are of the same shape, indicating that the average gra-

dients are the same for both times of development. The lateral spacing along the logarithm of relative exposure axis is a measure of the speed increase resulting from the increased development.

Consider two slightly different thicknesses in a specimen, one transmitting 25 percent more x-radiation than the other. The difference in log relative exposure between the two areas of the specimen ($\Delta \log E$) is 0.10. With a certain exposure time (Exposure 1) a radiograph developed for 5 minutes has a density of 0.36 (ΔD_5) between the two areas. On a radiograph given the same exposure but developed for 8 minutes, the density difference is 0.55 (ΔD_8). If, however, the exposure time is decreased to just compensate for the increased film speed at 8 minutes' development (Exposure 2) the log relative exposures corresponding to the two areas of the specimen are moved to the left, although the spacing between them (in terms of log exposure) remains constant. In this latter case, the density difference for 8 minutes' development is the same as that for 5 minutes, that is, 0.36. This, of course, is merely a further illustration of the fact that the density difference corresponding to a certain difference in relative x-ray exposure depends on the region of the characteristic curve where the exposures fall.

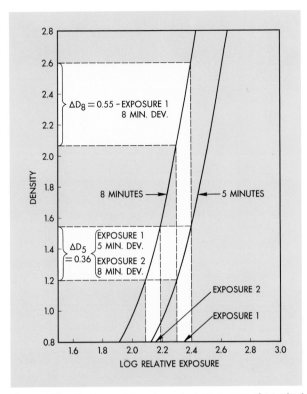

Figure 123—Portions of the characteristic curves of a typical industrial x-ray film developed for 5 and 8 minutes. Density differences corresponding to pairs of exposures differing by 25 percent are shown for each development time. (Scale of figure expanded from that in Figures 114, 115, 116 and 118 for ease of illustration.)

X-RAY SPECTRAL SENSITIVITY

As has been pointed out, the shape of the characteristic curve of an x-ray film is unaffected, for practical purposes, by the wavelength (energy) of the x-rays or gamma rays to which the film is exposed. However, the *sensitivity* of the film, in terms of the number of roentgens required to produce a given density, is strongly affected by the energy (wavelength) of the exposing radiation.

Figure 124 shows the number of roentgens needed to produce a density of 1.0 on a particular x-ray film under certain processing conditions. (Exposures were made without screens, either fluorescent or lead.)

The spectral sensitivity curves for all x-ray films have roughly the same general features as the curves shown in Figure 124. Details, among them the ratio of maximum to minimum sensitivity, differ from type to type of film, however.

The spectral sensitivity of a film, or differences in spectral sensitivity between two films, need rarely be considered in industrial radiography. Usually, such changes in sensitivity are automatically taken into account in the preparation of exposure charts (see page 50) and of tables of relative film speeds. The spectral sensitivity of a film is, of course, very important in radiation monitoring, because in this case an evaluation of the number of roentgens incident upon the film is required.

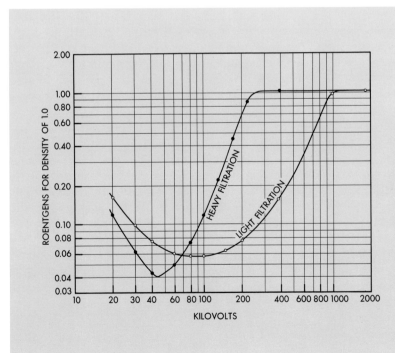

Figure 124—Typical x-ray spectral sensitivity curve of an x-ray film, showing number of roentgens required to produce a density of 1.0 for various radiation qualities. Other films will have curves of similar shape, but shifted up or down depending upon the properties of the film and the development technique used. (After Wilsey, Radiology, 56:229, 1951.)

*Net density—density above base density and fog.

143

GRAININESS

Graininess is defined as the visual impression of nonuniformity of density in a radiographic (or photographic) image. With fast films exposed to high-kilovoltage radiation, the graininess is easily apparent to the unaided vision; with slow films exposed to low-kilovoltage x-rays, moderate magnification may be needed to make it visible. In general, graininess increases with increasing film speed and with increasing energy of the radiation.

The "clumps" of developed silver which are responsible for the impression of graininess do not each arise from a single developed photographic grain. That this cannot be so can be seen from size consideration alone. The particle of black metallic silver arising from the development of a single photographic grain in an industrial x-ray film (see Figure 69, page 71) is rarely larger than 0.001 mm (0.00004 inch) and usually even less. This is far below the limits of unaided human vision.

Rather, the visual impression of graininess is caused by the random, statistical grouping of these individual silver particles. Each quantum (photon) of x-radiation or gamma radiation absorbed in the film emulsion exposes one or more of the tiny crystals of silver bromide of which the emulsion is composed (see Figure 67, page 71). These "absorption events" occur at random and even in a uniform x-ray beam, the number of absorption events will differ from one tiny area of the film to the next for purely statistical reasons. Thus, the exposed grains will be randomly distributed; that is, their numbers will have a statistical variation from one area to the next.

In understanding this effect, a simple analogy—a long sidewalk on a rainy day—will be of assistance. The sidewalk corresponds to the film and the raindrops to the x-ray photons absorbed in it. (Only those *absorbed* by the film are considered, because only those that are absorbed have a photographic effect.) First, consider what happens in a downpour so hard that there is an *average* of 10,000 drops per block or square of the sidewalk. It would not be expected, however, that each square would

receive precisely this average number of 10,000 drops. Since the raindrops fall at random, only a few or perhaps none of the squares would receive precisely 10,000 drops. Some would receive more than 10,000 and others less. In other words, the actual number of drops falling on any particular square will most likely differ from the average number of drops per square along the whole length of the sidewalk. The laws of statistics show that the *differences* between the actual numbers and the average number of drops can be described in terms of probability. If a large number of blocks is involved, the actual number of raindrops on 68 percent of the blocks will differ from the average by no more than 100 drops or ± 1 percent of the average.* The remaining 32 percent will differ by more than this number. Thus, the differences in "wetness" from one block of sidewalk to another will be small and probably unnoticeable.

This value of 100 drops holds only for an average of 10,000 drops per square. Now consider the same sidewalk in a light shower in which the average number of drops per square is only 100. The same statistical laws show that the deviation from the average number of drops will be 10 or ± 10 percent of the average. Thus, differences in wetness from one square to the next will be much more noticeable in a light shower (± 10 percent) than they are in a heavy downpour (± 1 percent).

Now we will consider these drops as x-ray photons absorbed in the film. With a very slow film, it might be necessary to have 10,000 photons absorbed in a small area to produce a density of, for example, 1.0. With an extremely fast film it might require only 100 photons in the same area to produce the same density of 1.0. When only a few photons are required to produce the density, the random positions of the absorption events become visible in the processed film as film graininess. On the other hand, the more x-ray photons required,

---

*A statistician refers to these values as the standard deviation (± 100 drops) or the relative deviation (± 1 percent) from the average. They are calculated by taking the square root of the average. Thus, if the average number is 10,000, as in this illustration, the deviation from the average is $\sqrt{10,000}$ or ± 100. In turn, 100 is 1 percent of the 10,000, so the relative deviation is ± 1 percent. Standard deviation is usually symbolized by the small Greek sigma (σ).

the less noticeable the graininess in the radiographic image, all else being equal.

It can now be seen how film speed governs film graininess. In general, the silver bromide crystals in a slow film are smaller than those in a fast film, and thus will produce less light-absorbing silver when they are exposed and developed. Yet, at low kilovoltages, one absorbed photon will expose one grain, of whatever size. Thus, more photons will have to be absorbed in the slower film than in the faster to result in a particular density. For the slower film, the situation will be closer to the "downpour" case in the analogy above and film graininess will be lower.

The increase in graininess of a particular film with increasing kilovoltage can also be understood on this basis. As pointed out on page 151, at low kilovoltages each absorbed photon exposes one photographic grain; at high kilovoltages one photon will expose several, or even many, grains. At high kilovoltages, then, fewer absorption events will be required to expose the number of grains required for a given density than at lower kilovoltages. The fewer absorption events, in turn, mean a greater relative deviation from the average, and hence greater graininess.

It should be pointed out that although this discussion is on the basis of direct x-ray exposures, it also applies to exposures with lead screens. On page 151 it was mentioned that the agent which actually exposes a grain is a high-speed electron arising from the absorption of an x- or gamma-ray photon. The silver bromide grain in a film cannot distinguish between an electron which arises from an absorption event within the film emulsion and one arising from the absorption of a similar photon in a lead screen.

The quantum mottle (see page 35) observed in radiographs made with fluorescent intensifying screens has a similar statistical origin. In this case, however, it is the numbers of photons absorbed in the screens which are of significance.

SIGNAL-TO-NOISE RATIO

The statistical reasoning used above has an application in considering the radiographic recording of small details. Again, the discussion will be in terms of direct x-ray exposures to industrial radiographic films, but the same principles apply to exposures with lead screens.

Because of the random spatial distribution of the photons absorbed in the film, the exposure to any small area of a film in a uniform beam of radiation is likely to differ by a small amount from the exposure to another area of the same size. This "noise" naturally interferes with the visibility of a small or faint detail, because the image of the detail may be lost in the density variations arising statistically.

It can be shown that for a given small area (A) on a uniformly exposed film, the noise (σ D)—that is, the standard deviation of density (see footnote, page 144)—is proportional to the square root of the exposure to the film.* For this discussion it will be convenient to state the exposure in terms of the number (N) of photons *absorbed* in the film within the small area, A, being considered. Those photons not absorbed have no photographic effect.

This can be stated mathematically:

$$\text{Noise} = \sigma\,D \propto \sqrt{N}$$

The "signal" transmitted to the film by a detail of area A in the object being radiographed is the difference (ΔN)† between the number of photons reaching the film through the detail and the average number reaching similar areas of film outside the detail image. This signal is proportional to the exposure or

$$\text{Signal} = \Delta N \propto N$$

The ratio of signal to noise has a profound bearing on the minimum size of detail that can be seen. It has been shown that for threshold visibility of detail, this signal-to-noise ratio must be at least 5.

An example can be given to illustrate how the information content of a radiograph depends upon exposure—that is, depends upon the number of x-ray photons involved in forming the image. Consider a penetrameter on a flat steel plate (Figure 125). The penetrameter has a single hole, the image of which is to be made visible in the radiograph. The area of the hole (A) will be the unit of area. Suppose that in a 0.01-second exposure, for example, an average of 100 of the photons per unit area (A) transmitted through the plate and penetrameter body interact with the film to form an image. In the area of the hole, however, assume that an average of 101 photons interact with the film in the same time. Thus the "signal" which it is hoped to detect—the excess of radiation through the hole above that in the background—is one photon. *But,*

---

*This is true for conditions under which each absorbed photon exposes one or more photographic grains—that is, when energy is not "wasted" on grains already exposed by a previous photon. This does not occur until a significant fraction of the total number of grains has been exposed. Most industrial x-ray films have such a large number of grains that wastage of energy from this cause occurs only at high densities.
†The Greek letter delta (Δ) has the significance of "difference in" or "change in."

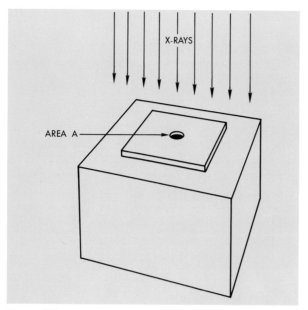

Figure 125

the standard deviation of the number of photons in the background is $\sqrt{100} = 10$. In this case, it would be impossible to detect the signal (the image of the penetrameter hole) because the "noise" is ten times as great as the signal.

It will now be interesting to see what happens when the number of photons that are absorbed in the film is increased. This can be done by increasing the exposure time above the 0.01 second originally assumed in this example. The table below shows the conditions for a series of increasing times. "Background" is the number of absorbed photons per "hole area" beneath plate and penetrameter; "signal" is the excess of absorbed photons through the hole.

Exposure Time (Seconds)	Background	Signal	Noise	Signal-Noise Ratio
0.01	100	1	$\sqrt{100}$	0.1
0.1	1,000	10	$\sqrt{1,000}$	0.316
1.0	10,000	100	$\sqrt{10,000}$	1
10.0	100,000	1,000	$\sqrt{100,000}$	3.16
30.0	300,000	3,000	$\sqrt{300,000}$	5.5
100.0	1,000,000	10,000	$\sqrt{1,000,000}$	10

Since threshold visibility of a detail is not achieved until the signal-to-noise ratio is at least 5, the penetrameter hole would not be visible until an exposure of 30 seconds (in this example) had been reached. If the exposure were extended to 100 seconds, visibility of the detail would be more certain, because the signal-to-noise ratio would be increased to 10.

Suppose, now, there were two conventional industrial radiographic films, one of which gave a density of 2.0 for 300,000 absorbed photons over the area A of the penetrameter hole (exposure of 30 seconds in the example used), and the other about one-third the speed, requiring 1,000,000 photons absorbed over the same area to give the same density. The penetrameter hole would be much more visible on the slower film, because of the better signal-to-noise ratio. Indeed, it can be said that the slower film gives a better image *because* it requires more radiation to produce the image. In many instances, with the exposure required for the slower film, a hole about one-third the area (about 0.55 the diameter) of the original hole would be at the threshold of visibility.

In the preceding discussion nothing has been said about film contrast. Actually, of course, *both* adequate film contrast and a sufficiently high signal-to-noise ratio are essential to the visibility of a particular detail in a radiograph.

If the densities involved fall on the toe of the characteristic curve where the film contrast is very low (see page 137), the image of the detail will be invisible to the observer, no matter what the signal-to-noise ratio might be. Further, suppose that an exposure were made, the densities of which fell on the shoulder of the characteristic curve for Film Z (Figure 116). Again, the image of the detail might be invisible to the observer, because of low film contrast. Decreasing the exposure time would make use of a steeper portion of the characteristic curve (giving higher film contrast) and produce a better radiograph. To sum up, increasing film contrast will increase the *ease* with which the image of a small detail can be visualized, provided always that the signal-to-noise ratio is above the required minimum.

Conversely, however, it can be stated that no increase in film contrast will make an image visible if the signal-to-noise ratio is inadequate. An increase in the film contrast with no change in exposure (that is, with no change in signal-to-noise ratio) will merely increase both the density variations due to noise and those due to the desired image, with no improvement in visibility of the detail. Suppose, for example, that a radiograph were made on Film Y at a density of 0.5 (Figure 116), and that the signal-to-noise ratio for a particular small penetrameter hole were very low—say 2. Changing to Film X, with all other conditions remaining the same, would give a higher film contrast. However, the signal-to-noise ratio would remain the same—namely, 2—and the image of the penetrameter hole would continue to be lost in the random density fluctuations of the background.

CHAPTER 18

The Photographic Latent Image

As shown in Figures 67 and 68, a photographic emulsion consists of a myriad of tiny crystals of silver halide—usually the bromide with a small quantity of iodide—dispersed in gelatin and coated on a support. The crystals—or photographic grains—respond as individual units to the successive actions of radiation and the photographic developer (Figure 69).

The *photographic latent image* may be defined as that radiation-induced change in a grain or crystal that renders the grain readily susceptible to the chemical action of a developer.

To discuss the latent image in the confines of a single chapter requires that only the basic concept be outlined. A discussion of the historical development of the subject and a consideration of most of the experimental evidence supporting these theories must be omitted because of lack of space.

It is interesting to note that throughout the greater part of the history of photography, the nature of the latent image was unknown or in considerable doubt. The first public announcement of Daguerre's process was made in 1839, but it was not until 1938 that a reasonably satisfactory and coherent theory of the formation of the photographic latent image was proposed. That theory has been undergoing refinement and modification ever since.

Some of the investigational difficulties arose because the formation of the latent image is a very subtle change in the silver halide grain. It involves the absorption of only one or a few photons of radiation and can therefore affect only a few atoms, out of some 10^9 or 10^{10} atoms in a typical photographic grain. The latent image cannot be detected by direct physical or analytical chemical means.

However, even during the time that the mechanism of formation of the latent image was a subject for speculation, a good deal was known about its physical nature. It was known, for example, that the latent image was localized at certain discrete sites on the silver halide grain. If a photographic emulsion is exposed to light, developed briefly, fixed, and then examined under a microscope (Figure 126), it can be seen that development (the reduction of silver halide to metallic silver) has begun at only one or a few places on the crystal. Since small amounts of silver sulfide on the surface of the grain were known to be necessary for a photographic material to have a high sensitivity, it seemed likely that the spots at which the latent image was localized were local concentrations of silver sulfide.

It was further known that the material of the latent image was, in all probability, silver. For one thing, chemical reactions that will oxidize silver will also destroy the latent image. For another, it is a common observation that photographic materials given prolonged exposure to light darken spontaneously, without the need for development. This darkening is known as the print-out image. The print-out image contains enough material to be identified chemically, and this material is metallic silver. By microscopic examination, the silver of the print-out image is discovered to be localized at

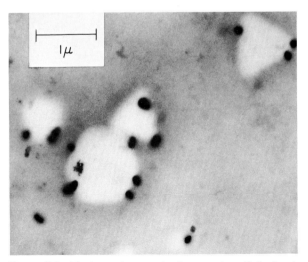

Figure 126—Electron micrograph of exposed, partially developed, and fixed grains, showing initiation of development at localized sites on the grains ($1\mu = 1$ micron $= 0.001$ mm).

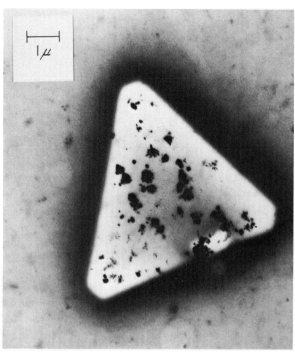

Figure 127—Electron micrograph of photolytic silver produced in a grain by very intense exposure to light.

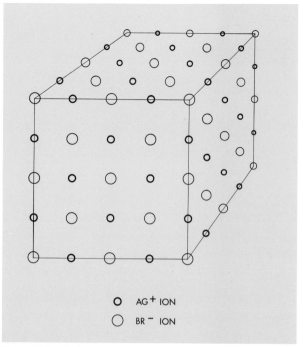

O AG⁺ ION

O BR⁻ ION

Figure 128—A silver bromide crystal is a rectangular array of silver (Ag+) and bromide (Br−) ions.

certain discrete areas of the grain (Figure 127), just as is the latent image.

Thus, the change that makes an exposed photographic grain capable of being transformed into metallic silver by the mild reducing action of a photographic developer is a concentration of silver atoms—probably only a few—at one or more discrete sites on the grain. Any theory of latent-image formation must account for the way that light photons absorbed at random within the grain can produce these isolated aggregates of silver atoms.

Most current theories of latent-image formation are modifications of the mechanism proposed by R. W. Gurney and N. F. Mott in 1938.

In order to understand the Gurney-Mott theory of the latent image, it is necessary to digress and consider the structure of crystals—in particular, the structure of silver bromide crystals.

When solid silver bromide is formed, as in the preparation of a photographic emulsion, the silver atoms each give up one orbital electron to a bromine atom. The silver atoms, lacking one *negative* charge, have an effective positive charge and are known as silver *ions* (Ag+). The bromine atoms, on the other hand, have gained an electron—a negative charge—and have become bromine ions (Br−). The "plus" and "minus" signs indicate, respectively, one fewer or one more electron than the number required for electrical neutrality of the atom.

A crystal of silver bromide is a regular cubical array of silver and bromide ions, as shown schematically in Figure 128. It should be emphasized that the "magnification" of Figure 128 is very great. An average grain in an industrial x-ray film may be about 0.00004 inch in diameter, yet will contain several billions of ions.

A crystal of silver bromide in a photographic emulsion is—fortunately—not perfect; a number of imperfections are always present. First, within the crystal, there are silver ions that do not occupy the "lattice positions" shown in Figure 128, but rather are in the spaces between. These are known as interstitial silver ions (Figure 129). The number of the interstitial silver ions is, of course, small compared to the total number of silver ions in the crystal. In addition, there are distortions of the uniform crystal structure. These may be "foreign" molecules, within or on the crystal, produced by reactions with the components of the gelatin, or distortions or dislocations of the regular array of ions shown in Figure 128. These may be classed together and called "latent-image sites."

The Gurney-Mott theory envisions latent-image formation as a two-stage process. It will be discussed first in terms of the formation of the latent image by light, and then the special considerations of direct x-ray or lead foil screen exposures will be covered.

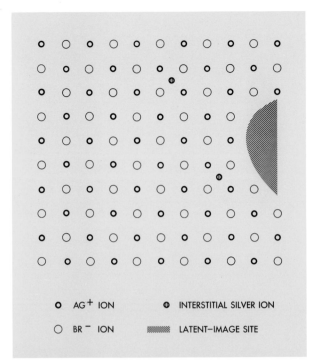

Figure 129—"Plan view" of a layer of ions of a crystal similar to that of Figure 128. A latent-image site is shown schematically, and two interstitial silver ions are indicated.

THE GURNEY-MOTT THEORY

When a photon of light of energy greater than a certain minimum value (that is, of wavelength less than a certain maximum) is absorbed in a silver bromide crystal, it releases an electron from a bromide (Br−) ion. The ion, having lost its excess negative charge, is changed to a bromine atom. The liberated electron is free to wander about the crystal (Figure 130). As it does, it may encounter a latent-image site and be "trapped" there, giving the latent-image site a negative electrical charge (Figure 130). This first stage of latent-image formation—involving as it does transfer of electrical charges by means of moving electrons—is the electronic conduction stage.

The negatively charged trap can then attract an interstitial silver ion because the silver ion is charged positively (Figure 130C). When such an interstitial ion reaches a negatively charged trap, its charge is counteracted, an atom of silver is deposited at the trap, and the trap is "reset" (Figure 130D). This second stage of the Gurney-Mott mechanism is termed the ionic condition stage, since electrical charge is transferred through the crystal by the movement of ions—that is, charged atoms. The whole cycle can recur several, or many, times at a single trap, each cycle involving absorp-

tion of one photon and addition of one silver atom to the aggregate. (Figure 130, E to H)

In other words, this aggregate of silver atoms *is* the latent image. The presence of these few atoms at a single latent-image site makes the whole grain susceptible to the reducing action of the developer. In the most sensitive emulsions, the number of silver atoms required may be less than ten.

The mark of the success of a theory is its ability to provide an understanding of previously inexplicable phenomena. The Gurney-Mott theory and those derived from it have been notably successful in explaining a number of photographic effects. One of these effects—reciprocity-law failure (see pages 48 and 140)—will be considered here as an illustration.

Low-intensity reciprocity-law failure (left branch of the curve in Figure 120) results from the fact that several atoms of silver are required to produce a stable latent image. A single atom of silver at a latent-image site (Figure 130D) is relatively unstable, breaking down rather easily into an electron and a positive silver ion. Thus, if there is a long interval between the formation of the first silver atom and the arrival of the second conduction electron (Figure 130E), the first silver atom may have broken down, with the net result that the energy of the light photon that produced it has been wasted. Therefore, increasing light intensity from very low to higher values increases the efficiency, as shown by the downward trend of the left-hand branch of the curve in Figure 120, as intensity increases.

High-intensity reciprocity-law failure (right branch of the curve of Figure 120) is frequently a consequence of the sluggishness of the ionic process in latent-image formation (Figure 130). According to the Gurney-Mott mechanism, a trapped electron must be neutralized by the movement of an interstitial silver ion to that spot (Figure 130D) before a second electron can be trapped there (Figure 130E); otherwise, the second electron is repelled and may be trapped elsewhere. Therefore, if electrons arrive at a particular sensitivity center faster than the ions can migrate to the center, some electrons are repelled, and the center does not build up with maximum efficiency. Electrons thus denied access to the same traps may be trapped at others, and the latent image silver therefore tends to be inefficiently divided among several latent-image sites. (This has been demonstrated by experiments which have shown that high-intensity exposure produces more latent image within the volume of the crystal than do either low- or

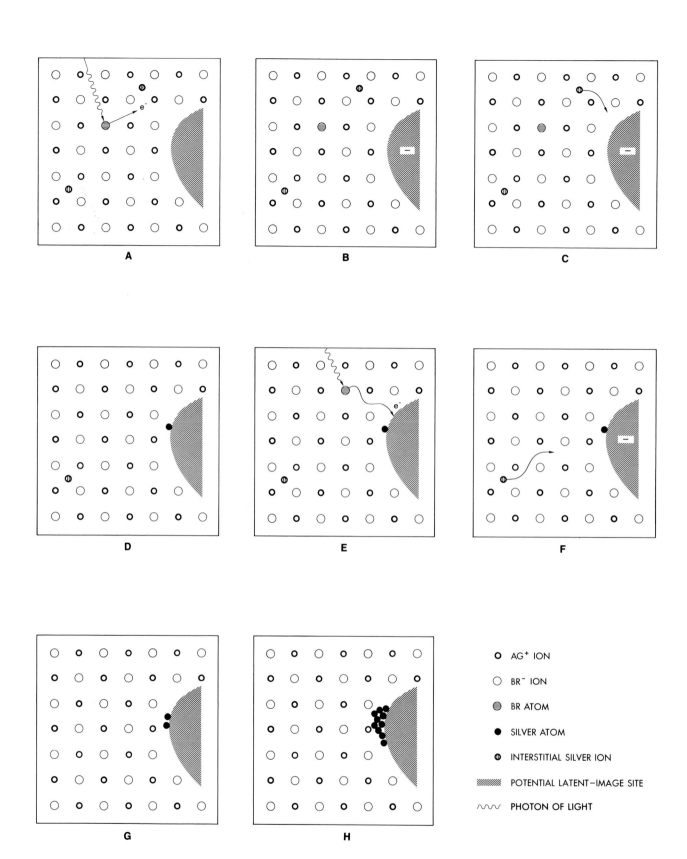

Figure 130—Stages in the development of the latent image according to the Gurney-Mott theory.

optimum-intensity exposures.) Thus, the resulting inefficiency in the use of the conduction electrons is responsible for the upward trend of the right-hand branch of the curve of Figure 120.

X-RAY LATENT IMAGE

In industrial radiography, the photographic effects of x-rays and gamma rays, rather than those of light, are of the greater interest.

At the outset it should be stated that the agent that actually exposes a photographic grain, that is, a silver bromide crystal in the emulsion, is not the x-ray photon itself, but rather the electrons—photoelectric and Compton—resulting from the absorption event. It is for this reason that direct x-ray exposures and lead foil screen exposures are similar and can be considered together.

The most striking differences between x-ray and visible-light exposures to grains arise from the difference in the amounts of energy involved. The absorption of a single photon of light transfers a very small amount of energy to the crystal. This is only enough energy to free a *single* electron from a bromide (Br$^-$) ion, and several successive light photons are required to render a single grain developable. The passage through a grain of an electron, arising from the absorption of an x-ray photon, can transmit hundreds of times more energy to the grain than does the absorption of a light photon. Even though this energy is used rather inefficiently, in general the amount is sufficient to render the grain traversed developable—that is, to produce within it, or on it, a stable latent image.

As a matter of fact, the photoelectric or Compton electron, resulting from absorption or interaction of a photon, can have a fairly long path in the emulsion and can render several or many grains developable. The number of grains exposed per photon interaction can vary from 1 grain for x-radiation of about 10 keV to possibly 50 or more grains for a 1 meV photon. However, for 1 meV and higher energy photons, there is a low probability of an interaction that transfers the total energy to grains in an emulsion. Most commonly, high photon energy is imparted to several electrons by successive Compton interactions. Also, high-energy electrons pass out of an emulsion before all of their energy is dissipated. For these reasons there are, on the average, 5 to 10 grains made developable per photon interaction at high energy.

For comparatively low values of exposure, each increment of exposure renders on the average the

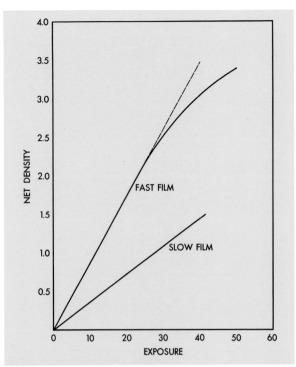

Figure 131—Typical net density versus exposure curves for direct x-ray exposures.

same number of grains developable which, in turn, means that a curve of *net* density versus exposure is a straight line passing through the origin (Figure 131). This curve departs significantly from linearity only when the exposure becomes so great that appreciable energy is wasted on grains that have already been exposed. For commercially available fine-grain x-ray films, for example, the density versus exposure curve may be essentially linear up to densities of 2.0 or even higher.

The fairly extensive straight-line relation between exposure and density is of considerable use in photographic monitoring of radiation, permitting a saving of time in the interpretation of densities observed on personnel monitoring films.

If the D versus E curves shown in Figure 131 are replotted as characteristic curves (D versus log E), both characteristic curves are the same shape (Figure 132) and are merely separated along the log exposure axis. This similarity in toe shape has been experimentally observed for conventional processing of many commercial photographic materials, both x-ray films and others.

Because a grain is completely exposed by the passage of an energetic electron, all x-ray exposures are, as far as the *individual* grain is concerned, extremely short. The actual time that an x-ray-induced electron is within a grain depends upon the electron velocity, the grain dimensions, and the

151

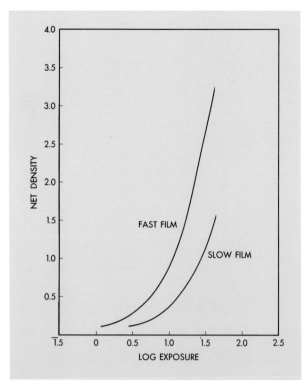

Figure 132—Characteristic curves plotted from data of Figure 131.

"squareness" of the hit. However, a time of the order of 10^{-13} second is representative. (This is in distinction to the case of light where the "exposure time" for a single grain is the interval between the arrival of the first photon and that of the last photon required to produce a stable latent image.)

The complete exposure of a grain by a single event and in a very short time implies that there should be no reciprocity-law failure for direct x-ray exposures or for exposures made with lead foil screens. The validity of this has been established for commercially available film and conventional processing over an extremely wide range of x-ray intensities. That films can satisfactorily integrate x-, gamma-, and beta-ray exposures delivered at a wide range of intensities is one of the advantages of film as a radiation dosimeter.

In the discussion on reciprocity-law failure it was pointed out that a very short, very high intensity exposure to light tends to produce latent images in the interior of the grain. Because x-ray exposures are also, in effect, very short, very high intensity exposures, they too tend to produce internal, as well as surface, latent images.

DEVELOPMENT

Many materials discolor on exposure to light—a pine board or the human skin, for examples—and

thus could conceivably be used to record images. However, most such systems react to exposure on a "1:1" basis, in that one photon of light results in the production of one altered molecule or atom. The process of development constitutes one of the major advantages of the silver halide system of photography. In this system, a few atoms of photolytically deposited silver can, by development, be made to trigger the subsequent chemical deposition of some 10^9 or 10^{10} additional silver atoms, resulting in an amplification factor of the order of 10^9 or greater. The amplification process can be performed at a time, and to a degree, convenient to the user and, with sufficient care, can be uniform and reproducible enough for the purposes of quantitative measurements of radiation.

Development is essentially a chemical reduction in which silver halide is reduced or converted to metallic silver. In order to retain the photographic image, however, the reaction must be limited largely to those grains that contain a latent image, that is, to those grains that have received more than a certain minimum exposure to radiation. Compounds that can be used as photographic developing agents, therefore, are limited to those in which the reduction of silver halide to metallic silver is catalyzed (or speeded up) by the presence of the metallic silver of the latent image. Those compounds that reduce silver halide in the absence of a catalytic effect by the latent image are not suitable developing agents because they produce a uniform overall density on the processed film.

Many practical developing agents are relatively simple organic compounds (Figure 133) and, as shown, their activity is strongly dependent on molecular structure as well as on composition. There exist empirical rules by which the developing activity of a particular compound may often be predicted from a knowledge of its structure.

The simplest concept of the role of the latent image in development is that it acts merely as an electron-conducting bridge by which electrons from the developing agent can reach the silver ions on the interior face of the latent image. Experiment has shown that this simple concept is inadequate to explain the phenomena encountered in practical photographic development. Adsorption of the developing agent to the silver halide or at the silver-silver halide interface has been shown to be very important in determining the rate of direct, or chemical, development by most developing agents. The rate of development by hydroquinone (Figure 133), for example, appears to be relatively independent of the area of the silver surface and

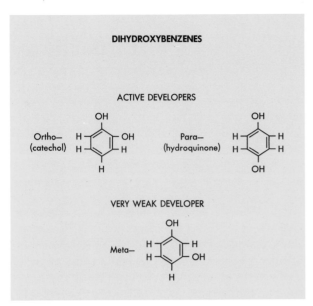

Figure 133—Configurations of dihydroxybenzene, showing how developer properties depend upon structure.

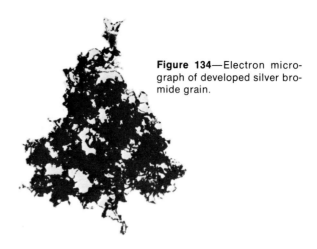

Figure 134—Electron micrograph of developed silver bromide grain.

instead to be governed by the extent of the silver-silver halide interface.

The exact mechanisms by which a developing agent acts are relatively complicated, and research on the subject is very active.

The broad outlines, however, are relatively clear. A molecule of a developing agent can easily give an electron to an exposed silver bromide grain (that is, to one that carries a latent image), but not to an unexposed grain. This electron can combine with a silver (Ag^+) ion of the crystal, neutralizing the positive charge and producing an atom of silver. The process can be repeated many times until all the billions of silver ions in a photographic grain have been turned into metallic silver.

The development process has both similarities to, and differences from, the process of latent-image formation. Both involve the union of a silver ion and an electron to produce an atom of metallic silver. In latent-image formation, the electron is freed by the action of radiation and combines with an interstitial silver ion. In the development process, the electrons are supplied by a chemical electron-donor and combine with the silver ions of the crystal lattice.

The physical shape of the developed silver need have little relation to the shape of the silver halide grain from which it was derived. Very often the metallic silver has a tangled, filamentary form, the outer boundaries of which can extend far beyond the limits of the original silver halide grain (Figure 134). The mechanism by which these filaments are formed is still in doubt although it is probably associated with that by which filamentary silver can be produced by vacuum deposition of the silver atoms from the vapor phase onto suitable nuclei.

The discussion of development has thus far been limited to the action of the developing agent alone. However, a practical photographic developer solution consists of much more than a mere water solution of a developing agent. The function of the other common components of a practical developer will be briefly discussed.

An alkali The activity of developing agents depends on the alkalinity of the solution. The alkali should also have a strong buffering action to counteract the liberation of hydrogen ions—that is, a tendency toward acidity—that accompanies the development process. Common alkalis are sodium hydroxide, sodium carbonate, and certain borates.

A preservative This is usually a sulfite. One of its chief functions is to protect the developing agent from oxidation by air. It destroys certain reaction products of the oxidation of the developing agent which tend to catalyze the oxidation reaction. Sulfite also reacts with the reaction products of the development process itself, thus tending to maintain the development rate and to prevent staining of the photographic layer.

A restrainer A bromide, usually potassium bromide, is a common restrainer or antifoggant. Bromide ions decrease the possible concentration of silver ions in solution (by the common-ion effect) and also, by being adsorbed to the surface of the silver bromide grain, protect unexposed grains from the action of the developer. Both of these actions tend to reduce the formation of fog.

Commercial developers often contain other materials in addition to those listed above. An example would be the hardeners usually used in developers for automatic processors.

153

CHAPTER 19

Protection

One of the most important considerations in the x-ray or gamma-ray laboratory is the provision and exercise of adequate safeguards for the personnel. Only the general principles of the necessary protective precautions are discussed in the following.

For details, pertinent National Bureau of Standards Handbooks, the publications of the National Council on Radiation Protection and Measurement, the Atomic Energy Commission regulations, and the state and local codes should be obtained and studied carefully. It is essential that new installations be constructed in compliance with the provisions of applicable codes and that existing installations be checked to make certain that they meet all requirements. It is desirable, and under some circumstances obligatory, to have a qualified radiation expert examine the radiation installation and protective measures. In addition, some state and local codes require that radiation-producing equipment be registered.

Any of the body tissues may be injured by excessive exposure to x-rays or gamma rays—the blood, the lens of the eye, and some internal organs being particularly sensitive. Unless exposure to x-rays or gamma rays is kept at a minimum, the cumulative effect may cause injury to the body, and it is essential that workers in the radiographic department be adequately protected against radiation at all times. Furthermore, protective measures should be so arranged that persons in nearby areas are also safe. Precautions should be particularly observed when radiography is done in the work areas of the shop rather than in a specially constructed department.

PROTECTION AGAINST X-RAYS

Exposure may be caused by the direct beam from the x-ray tube target or by scattered radiation arising from objects in the direct beam. Therefore, while exposures are being made, operators should always be protected by sufficient lead, or its equivalent, shielding them from the x-ray beam, the part being radiographed, and any other matter exposed to the x-rays.

Protection can be provided in a number of ways, depending upon the x-ray installation and the use to which it is put. Whenever possible, protective measures should be built in as permanent features of the installation. Preferably, the x-ray generator and the work should be enclosed in a room or cabinet, with the necessary protection incorporated in the walls.

The common method is to locate the x-ray tube within a room completely lined with lead of a sufficient thickness to provide adequate protection. All the controls are located outside the room.

In the placing of equipment and the design of protective enclosures, certain principles must be kept in mind. Careful application of these principles adds to the safety of the personnel, and may decrease cost.

Both safety and economy benefit if the amount of radiation that must be absorbed in the outside wall of the enclosure is kept to a minimum. To this end, the distance from the x-ray tube target to any occupied space should be as great as possible. Further, if the nature of the work permits, the direct beam should never be pointed in the direction of occupied areas.

Ideally, the lead housing around the x-ray tube should protect against all primary radiation except the useful beam, although this is not always feasible in practice. The useful beam itself should be limited in cross section by the use of cones, diaphragms or other devices.

If there are parts of the x-ray room that, because of the design of the equipment, can never be exposed to direct radiation, certain economies in the installation of protective material are possible.

Where *only* scattered radiation can reach a protective wall, less protection is necessary, since the intensity of the scattered radiation is much lower than that of the primary. When advantage is taken of these economies, great care must be exercised in rearranging equipment, lest it become possible to direct the full intensity of the x-ray beam against a wall which contains protection against the *scattered radiation only*.

Where large numbers of relatively small parts are inspected, the protection may be in a more compact form such as a lead-lined hood surrounding the x-ray tube, the specimens, and the film holder, completely enclosing them for the duration of the exposure. When the exposure is completed, the hood is opened to allow the removal of the radiographed parts and the placement of a new batch. The electrical controls are interlocked in order that the x-ray tube cannot be turned on until the hood is closed.

The protective material, usually lead, in the walls of the enclosure, whether it be a room or a cabinet, should be of sufficient thickness to reduce the exposure in all occupied areas to the lowest value possible or economically feasible. Under no circumstances should the exposure to personnel exceed that permitted by the pertinent regulations, and a lower level than this should be sought whenever possible.

In some cases, it may be possible for the personnel of an x-ray department, or other employees, to be exposed to the radiation from more than one x-ray machine. In such cases, the amount of protection must be increased to a point where the *total* exposure in any occupied area is within the prescribed radiation limits.

If the object being radiographed is too large or heavy to be brought to the x-ray machine, the radiography must be done in the shop. Under such conditions, special precautions are necessary. These include a completely lead-lined booth large enough to accommodate the x-ray machine controls, the operator, and his assistants. The booth may be completely enclosed, or open on one side. In any event, the exposure within it should be very carefully measured. Lead cones on the x-ray machine should be used to confine the x-ray beam to a certain direction and to the minimum angle that can be used. Portable screens should be provided to protect workers nearby. Guard rails or ropes and warnings should be used to keep others at a safe distance.

In field radiography, protection is usually obtained by distance. Care should be taken to see that all personnel are far enough away from the radiation source to insure safety.

MATERIAL AND CONSTRUCTION FOR PROTECTION AGAINST X-RAYS

Lead is the most common material used to provide protection against x-rays. It combines high protective efficiency with low cost and easy availability. In most cases, recommendations on protective measures are given in terms of lead thicknesses.

When using lead for protection, care must be taken to avoid any leaks in the shielding. This means that adjacent lead sheets should be overlapped, not merely butted, even if the sheets are to be burned together throughout the whole length of the joint. The heads of any nails or screws that pass through the lead should be carefully covered with lead. Extra precautions should be taken at those points where water pipes, electrical conduits, or ventilating ducts pass through the walls of the x-ray room. For small conduits and pipes, it is usually sufficient to provide a lead sheathing around the pipe for some distance on one side of the lead protective barrier in the wall. This sheath should be continuous and be very carefully joined, by a burned joint, to the lead in the wall. Better protection is afforded by having a right-angle bend in the pipe either inside or outside the x-ray room and covering it with a lead sheath to a point well beyond the right-angle bend. The sheath should be carefully joined to the lead in the wall. In the case of a large opening for ventilation, lead baffles will stop x-rays, but permit the passage of air. When a large ventilating duct is brought into the x-ray room, two right-angled bends covered with lead will prevent the escape of x-rays.

If the x-ray room is on the lowest floor of a building, the floor of the room need not be completely protected. The lead protection in the wall should not stop at the floor level, however. An "apron" of lead, continuous with the protection within the wall, should be placed in the floor, extending inward from all four walls. The purpose of this apron is to prevent x-rays from escaping from the room by penetrating the floor and then scattering upward outside the protective barrier. An alternative is to extend the lead protection in the walls downward for some distance below floor level. The same considerations apply to the ceiling if the x-ray room is located on the top floor of a building. Of course, if there is occupied space above or below the x-ray room, the ceiling or floor of the x-ray room must have full radiation protection over its whole area.

Although lead is the most common material for x-ray protection, other materials may be used. In particular, structural walls of concrete or brick may afford considerable protection and may reduce the thickness, and therefore the cost, of the lead required. Above 400 kV, concrete is most used as protective material. The thicknesses of lead required at these potentials are so great that fastening the lead to the walls becomes a serious problem, and concrete is often used because of the ease of construction. In new construction, the use of concrete may have economic advantages even for protection against radiations generated at kilovoltages well below 400. State and local codes should be examined, and any installations checked for compliance with their requirements.

PROTECTION AGAINST GAMMA RAYS

Most gamma-ray emitters used in industry are artificial radioactive isotopes; the procurement, use, handling, storage, and the like are controlled directly by the United States Atomic Energy Commission, or indirectly by state radiation control laws approved by the Atomic Energy Commission. It is essential, therefore, that these codes be followed rigorously.

Gamma rays may be very penetrating. For instance, one-half inch of lead reduces the intensity of the gamma rays of cobalt 60 only about 50 percent. This makes the problems of protection somewhat different from those encountered in protection against moderate-voltage x-rays. In general, it is not feasible to provide safety from gamma rays solely by means of a protective barrier. Therefore, distance or a combination of distance and protective material is usually required. When radioactive materials are not in use, protection is usually obtained by keeping them in thick lead containers, because in this case the total amount of lead needed is not great.

Because of the great thicknesses of protective materials required for shielding some gamma-ray sources, distance is the most economical method of protection while the source is in use. A danger zone should be roped off around the location of the radioactive material, and personnel should be forbidden to enter this zone except to put the source in position or return it to its safe. Suitable conspicuous signs should be provided to warn away the casual passersby. Tables are available which give data for calculating the distances from various amounts of radioactive material at which a radiation hazard exists.

It must be kept in mind that the presence of a large mass of scattering material, for example, a wall, will materially increase the gamma-ray exposure. This increase may be as much as 50 percent of the exposure as calculated without the presence of scattering material. Thus, to be sure that the radiation protection is adequate, factors other than distance must be kept in mind when considering personnel protection from gamma rays.

Precautions must be taken in shipping radioactive materials, not only to protect those who handle them in transit but also to prevent the fogging of photographic materials that may be transported in the same vehicle. The Interstate Commerce Commission has established regulations governing the rail shipments of radioactive isotopes. These provide in part: that the package shall have such internal shielding that the gamma radiation does not exceed 10 milliroentgens (mR) per hour at a distance of 3 feet from the outside of the container; that the radiation intensity at any readily accessible surface of the package shall not be over 200 mR per hour; that the package shall carry a prescribed label; and that shipping conditions be such that unprocessed photographic film traveling in the same vehicle shall be protected from damage throughout the transit period.

Packages meeting these requirements often consist of a central lead container for the radioactive isotope surrounded by a wooden or other box of such dimensions that the radiation at any readily accessible surface is less than 200 mR per hour. It is advisable to preserve the original package in case it is again necessary to ship the source.

INDEX

effect on:
  characteristic curve *139*
  radiographic quality *65, 139*
  subject contrast *28, 65-66*
  x-ray absorption *25-26*
  x-ray quality and intensity *22-23*

L

laminagraphy *111*
lamps, safelight *101*
latent image, photographic *147-152*
law, inverse square *23-24, 46*
law, reciprocity *46, 48, 152*
  failure of *48, 137, 140-141, 149*
laws, transformer *9*
lead, radiation protection with *155*
  radiographic equivalence factors for *25*
lead oxide screens, see screens, lead
lead screens, see screens, lead
lenses for enlarging microradiographs *125*
light lock *101*
linear accelerator *10-11*
line focus *8*
loading bench *97*
localization, depth, of defects *114-116*
logarithms *48-49*

M

magnesium, radiographic equivalence factors for *25*
magnification, geometric *15-20, 118, 125*
mantissa *49*
manual processing *79-85*
marks, pressure *37, 73*
  static *37, 73*
masks *39*
materials, radioactive
    (see also gamma-ray sources),
    radiography of *111-112*
maze *99*
metallic foil screens, see screens, lead
metallic shot for masking *40*
micrography, x-ray *123-127*
microradiographs, enlargement of *125-127*
microradiography *123-127*
  "reflection" *128*
milliamperage, control of *7*
  effect on x-ray intensity *22*
milliamperage (source strength)-distance-time
    relationships *46-48*
million-volt radiography *40, 45*

"miniature radiography" *123*
motion pictures, x-ray *123*
motion unsharpness in "in-motion"
    radiography *109-110*
mottle, due to lead backing in exposure holders *31*
    quantum (screen) *35, 67, 145*
    x-ray diffraction *43-44*
multi-thickness specimens, exposures for *60*

N

neutron radiography *114, 118*
"noise" in radiographs *145*
nomogram methods for exposure calculations *56-57*
number, atomic *7, 25*

O

orthogonal projection *110*
output, gamma-ray sources *11-13, 23*
output, x-ray, see x-ray intensity
overlays, for exposure calculation *55-56*
overreplenishment, effects of *87*

P

packaging, film *72-73*
paper, photographic, for enlarging
    microradiographs *126*
paper radiographs *132-135*
    comparing densities—paper and film *135*
    density—transmission vs reflection *134*
    density measurement *134*
    interpreting *135*
    viewing *134*
paper radiography *132-135*
    advantages of *132*
    applications for *132*
    factors affecting *132*
    exposure techniques *132*
    processing techniques *132-134*
      stabilization *132-133*
      automatic *132, 133*
      manual *134*
parallax method of depth localization *115*
peak kilovolts *9*
penetrameters *67-70*
penetration (of x-ray beam), see quality of
    radiation; kilovoltage

Q

R